Springer Series in Reliability Engineering

Series Editor

Hoang Pham

For further volumes:
http://www.springer.com/series/6917

Genserik Reniers · Yulia Pavlova

Using Game Theory to Improve Safety within Chemical Industrial Parks

Springer

Genserik Reniers
Antwerp Research Group
University of Antwerp
Antwerp
Belgium

Yulia Pavlova
Economic Research Unit
MTT Agrifood Research Finland
Helsinki
Finland

ISSN 1614-7839
ISBN 978-1-4471-5051-0 ISBN 978-1-4471-5052-7 (eBook)
DOI 10.1007/978-1-4471-5052-7
Springer London Heidelberg New York Dordrecht

Library of Congress Control Number: 2013933592

Printed on acid-free paper

Springer is part of Springer Science+Business Media (www.springer.com)

Preface

Game theory is more than a mathematical gadget for the chemical industry

The chemical industry covers several industrial sectors, including production of chemicals and the pharmaceutical industry but also the industries manufacturing, e.g., paints, varnishes, soaps, detergents, etc. The common element in these sectors is the handling of chemicals on an industrial scale. Companies storing and transporting hazardous chemicals may also be considered part of this industry.

In the chemical industry, economies of scope, environmental factors, social motives, and legal requirements often force companies to physically 'cluster'. Therefore, chemical plants are most often located in groups in so-called 'chemical industrial parks' or 'chemical clusters' and are rarely located separately. As Fortis and Maggioni (Curzio 2002) state, firms decide to settle in a cluster on the basis of the expected profitability of being located there. This profitability depends on geographical and agglomeration benefits, obtained as the difference between gross location-related benefits and costs. As the number of corporations located in an industrial cluster increases, gross benefits increase due to productive specialization, scientific, technical and economic spillovers, reduction in both transport and transaction costs, increases in the quality of the local pool of skilled labor force, etc. To maximize the clustering gross benefits, chemical organizations have a long tradition of collaborating on many different fronts. Their cooperative strategies offer significant advantages for plants that are lacking in particular competencies or resources to secure these through links with other firms possessing complementary skills or assets. They also offer opportunities for mutual synergy and learning (Reniers and Amyotte 2012). Hence, all over the world, chemical industrial parks bringing physically together a variety of individual chemical plants, are present.

In the case of chemical enterprises, clustering not only implies profit opportunities and economic benefits of scale. A chemical cluster has a very high responsibility toward maintaining safety standards in the urban surroundings as well. Each additional chemical plant entering a chemical cluster might decrease the average safety standing of the area.

Companies in chemical clusters are thus not merely linked via technological spillovers, logistics advantages, service agreements, and so on. They are related through the responsibility of gaining and sustaining safety standards in the entire industrial park as well. However, safety matters are often subject to a high degree of confidentiality. Therefore, safety cooperation strategies are difficult to establish. Nonetheless, since clustered chemical corporations are bonded by the responsibility to keep the industrial area as a whole as safe as possible, individual plants situated next to one another should develop a safety cooperation strategy.

In summary, strategic multi-organizational collaboration can lead to significant advantages in chemical industrial clusters, also—and especially—in the field of safety. Determining how best to enhance collaboration between different chemical companies to ensure industrial cluster safety, is thus a topic of great concern. Furthermore, it is natural to turn to game theory as a mathematical tool for ideas and approaches to help optimize collaborative situations within a multi-decision maker context (such as a chemical cluster).

The focus of game theory is interdependence, situations in which an entire group of decision makers (e.g., a cluster of plants) is affected by the choices made by every individual decision maker within that group (e.g., an individual chemical plant). In such an interlinked situation, the interesting questions include for example, what will each individual decision maker guess about the others' choices, what action will each decision maker take, what is the outcome of these actions, and how can the outcome be optimized, does it make a difference if there is a supra-plant body to give direction to individual decisions, in what way can decisions be influenced, etc.

This book therefore presents the main ideas of game theory and how it can be employed within chemical industrial parks to enhance safety cooperation between the chemical plants belonging to a cluster. No specific mathematical knowledge is needed to understand the theory expounded in this book. Nonetheless, all concepts are defined precisely, and logical reasoning is used throughout. We are furthermore convinced that the only way to appreciate the theory is to see it in action. Therefore, we included practical and illustrative examples throughout the different chapters. The reader will clearly notice that the use of game theory in a chemical industrial park to increase safety, is very promising. Game theory is thus more than merely a mathematical gadget for the chemical industry.

References

Curzio AQ (2002) Dynamics and models in theory and practice. Physica-Verlag, Heidelberg

Reniers GLL, Amyotte P (2012) Prevention in the chemical and process industries: future directions. J Loss Prev Process Ind 25:227–231

Contents

Abbreviations

ALARA	As Low As Reasonably Achievable
BNE	Bayesian Nash Equilibrium
CEO	Chief Executive Officer
CIA	Cooperative Incentives Approach
D	Down
DPA	Delta Process Academy
EFQM	European Federation for Quality Management
ERM	Enterprise Risk Management
HSE	Health, Safety, and Environment
I	Invest
IDEAL	Improvement Diamond for Excellence Achievement and Leadership in Safety and Security-model
ISN	Industrial Symbiosis Networks
ISO	International Standardization Organization
KM	Knowledge Management
L	Left
MPC	Multi-Plant Council
M-PSMS	Multi-Plant Safety Management System
NAT	Normal Accident Theory
NE	Nash Equilibrium
NI	Not Invest
OHSAS	Occupational Health & Safety Assessment Series
ORDER	Optimizing the Risk Decision and Expertise Rad-model
P2T	People, Procedures, Technology model
PD	Prisoner's Dilemma game
PDCA	Plan, Do, Check, Act-model
PDRC	risk Policy, Decision making, Risk, organization Culture-model
R	Right
RM	Risk Management
SH	Stag Hunt game
SMS	Safety Management System
TISC	Tipping Inducing Sub-Cluster
U	Up

Chapter 1
Introduction: Why a Book on Game Theory for Safety Within the Chemical Industry?

1.1 Objectives and Prospective Readership

This book is not a book purely on game theory, nor is it a book purely discussing safety principles within the chemical industry. The book is about both these research domains. However, it was our ambition that this publication would not merely be a review of these two very important academic and industrial areas of interest. On the contrary, we envisioned a book to discuss the potential use and integration of game-theoretical (mathematical) concepts within the chemical industrial sector and applied to safety management, from a theoretical as well as a practical point of view. This vision is not easy to achieve. Mathematical rigor and user-friendliness (/legibility) as well as managerial hands-on knowledge and practitioner's easy-to-use demands had to be combined for our vision to turn into a reality. This translates the book's readership possibly to be as diverse as cooperative and noncooperative game theory researchers, safety managers of chemical companies, CEOs of plants situated in industrial clusters, academic safety researchers, applied economic sciences researchers, consultants, etc. Hence, satisfactory background information about both research disciplines needs to be given for fulfilling the expectations of such a broad readership, as well as straightforward enlightenment on integration aspects, requirements, and results.

Game theory is the study of how players should rationally play games. In the context of game theory, a game can be described (Straffin 1993) as a situation in which at least two players are involved (which may be individuals, but also countries, banks, biological species, chemical companies, etc.), whereby each player has a number of possible courses of action (called 'strategies') which he may choose to follow. The strategies chosen by each player then settle the outcome of the game, which reflects a certain (numerical) payoff to every player. Of course, each player, if rational, would like the game to end in an outcome which gives him as large a payoff as possible. However, in game theory, the outcome of a game depends on the strategic choices of all players. Hence, every player—through his choice of strategy—only partially has control over the final outcome. The possible outcomes of a game are thus characterized by complex multiplayer decision-

G. Reniers and Y. Pavlova, *Using Game Theory to Improve Safety within Chemical Industrial Parks*, Springer Series in Reliability Engineering,
DOI: 10.1007/978-1-4471-5052-7_1, © Springer-Verlag London 2013

making processes which can be mathematically presented and solved. Since the theories and game-theoretical concepts developed and derived in this work, as well as their (possible) applications within the chemical industry, may be generalized to other industrial sectors, the book may be of interest to game theory researchers in general.

Game theory can also be described as the science of strategic decision making, helping to understand the various decisions made in the course of competition and collaboration between players (for example corporations). Since game theory can predict the possible outcomes of complex decision-making processes, it is ever more becoming an essential tool for managers wishing to improve the results of their strategic choices. Nonetheless, the book needs to offer easy-to-use instruments, tools, and theories, if it is to be used by corporate management. The book is therefore accessibly written, explaining in as simple terms as possible, the underlying mathematics. Clear examples and helpful diagrams are used throughout.

This book aims at helping managers to expand the conceptual cross-plant safety framework in which they operate. The objectives of this book are to resolve practical cooperation difficulties, to find new solutions to old problems by giving practitioners a deeper understanding of the nature of safety incentives/taxes and cooperation, and to offer an alternative viewpoint on cross-fences safety problems, which may or may not yield solutions, but which will certainly lead to an increased insight into the objective nature of strategic decision making within chemical industrial parks.

The book is organized as follows. In Chap. 1 the reader is introduced to the topic of safety within the chemical industry, which continues further in Chap. 2. Chapter 1 sets out a target to discuss potential use of game theory and integration of game-theoretic modeling into safety management. To give the reader an overview of strategic decision-making techniques, Chap. 3 discusses the main features of game-theoretic modeling, equilibrium concepts, and information settings. In Chap. 4 a game-theoretic model for cross-plant prevention in a chemical industrial cluster is introduced. Based on this model, an algorithm to enhance safety collaboration within chemical clusters is proposed. Chapter 7 contains interpretations of derived game-theoretic analysis for managers. The book contains essential methodology descriptions, modeling applications, and illustrative examples that answer most up-to-date challenges in achieving improvements of safety within chemical industrial clusters (or, in other words, chemical industrial parks).

Adequate (efficient and effective) safety management requires managers to be well-informed, multidisciplinary, and highly flexible in their approach to problem solving and decision making. Chemical organizations are mostly situated in so-called chemical industrial parks, being increasingly complex areas with respect to cross-plant safety practices (amongst others). Chemical companies can therefore no longer afford to live in isolation from the safety expectations of their neighbors, or indeed the larger chemical industrial cluster in which they are situated and operate. The next section tackles the description and definitions of chemical clusters or parts thereof and the safety management difficulties associated with such multi-plant areas.

1.2 Chemical Industrial Parks

1.2.1 Different Levels for Managing Safety Within a Chemical Industrial Area

As Fortis and Maggioni (Curzio 2002) state, firms decide to settle in a cluster on the basis of the expected profitability of being located there. As the number of corporations located in an industrial cluster increases, benefits increase due to productive specialization, scientific, technical and economic spillovers, reduction in both transport and transaction costs, increases in the quality of the local pool of skilled labor force, etc. This observation also explains why chemical plants form chemical clusters. However, in the case of chemical enterprises, clustering not only implies profit opportunities and economic benefits of scale. A chemical cluster has a very high responsibility toward maintaining safety standards in the surroundings as well. Each additional chemical plant entering a chemical cluster might decrease the average safety standing of the cluster and, by extension, of the industrial and urban area.

Companies in chemical clusters are thus not merely linked via technological spillovers, logistics advantages, and so on. They are related through the responsibility of gaining and sustaining safety standards in the entire cluster as well.

To maximize the clustering benefits, chemical organizations have a long tradition of collaborating on many different fronts. Their cooperative strategies offer significant advantages for plants that are lacking in particular competencies or resources to secure these through links with other firms possessing complementary skills or assets. They also offer opportunities for mutual synergy and learning. However, as Child et al. (2005) explain, the organizational cultures can have a significant impact upon the implementation of cooperative strategies. Cultures can create serious barriers to collaboration between organizations, and yet, at the same time, the knowledge embodied in cultures can provide a valuable resource for cooperation strategies. It is widely accepted that culture is something that is learned. It is not inherited like human nature or, at least partially, personality. This has a very important practical implication for cooperation between the members of organizations coming from different organizational cultural environments. For it means that despite the way a certain company culture is absorbed by a plant's employees, it may always be possible to learn further cultural attributes through experience or training. In terms of safety topics, in particular, the latter observation is interesting. Safety matters are often subject to a high degree of confidentiality. Therefore, safety cooperation strategies are difficult to establish. However, since clustered chemical corporations are bonded by the responsibility to keep the industrial area *as a whole* as safe as possible, individual plants situated next to one another should develop a safety cooperation strategy, bringing diverse safety cultures closer together and investing in cross-plant safety measures and management practices. Avoiding incidents and accidents and their associated direct and indirect costs can lead to substantial hypothetical benefits, especially in case of averting major accidents such as those affecting different plants at once. Although the potential benefits are obvious, developing strategic cooperative investment approaches between different chemical plants is not at all straightforward or easy.

How to enhance cooperation between chemical plants in a chemical industrial park is a very complex question that game theory can answer through developing and solving mathematical models and is treated and explained in this book in Chaps. 3–7.

Managing safety within chemical plants requires a system of structures, responsibilities, procedures, and the availability of appropriate resources and technological know-how. Reniers (2010a) indicates that safety in a chemical surrounding can and should be managed at different levels: at installation, at factory (or business unit), at plant (or company, enterprise), and at multi-plant levels. Figure 1.1 illustrates the difference between the different mentioned terms. Plant (*e*) for example consists of only one installation within a single factory. Plant (*e*) forms a multi-plant area with plant (*c*) and plant (*f*), all three plants participating in the same multi-plant safety initiative. In this illustrative example, plant (*d*) chooses not to

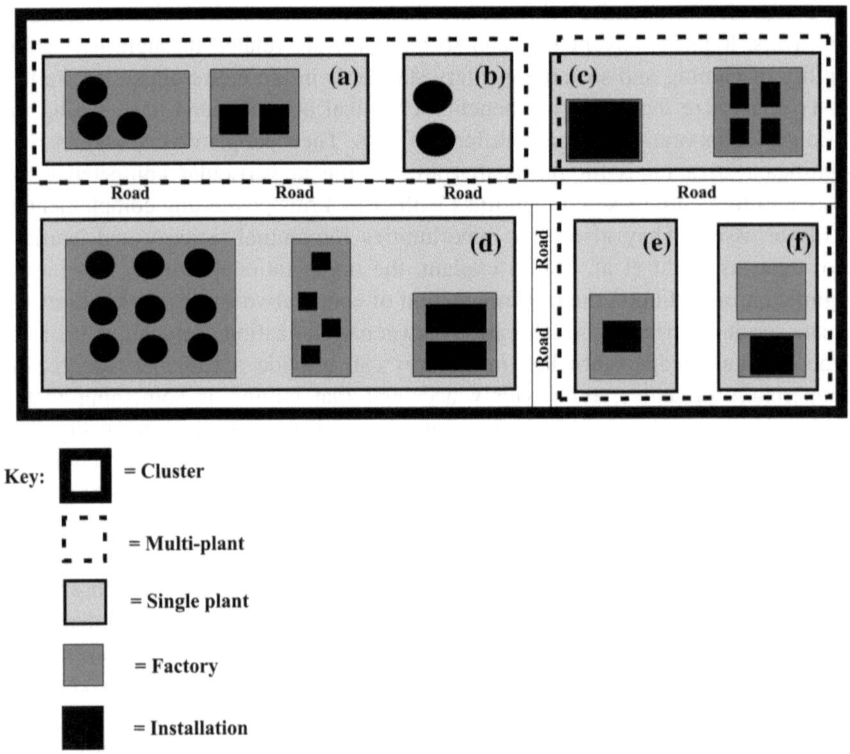

Fig. 1.1 Different levels within chemical clusters: installation, factory, plant, multi-plant, and cluster. T. Meyer, G. Reniers, Engineering risk management, 2013. Used here with kind permission © De Gruyter GmbH

collaborate intensively on a supra-plant level, whereas plants (*a*) and (*b*) form their own small-scale multi-plant safety initiative.

Factory-level safety includes topics such as working procedures, work packages, installation-specific training, personal protective equipment, quality inspection, etc. Plant-level safety includes defining acceptable safety risk levels and documenting plant-specific guidelines for implementing and achieving these levels for every facility situated on the premises of the plant. It is the current industrial practice to draft a plant's safety management system to meet these goals (Reniers 2010a). Multi-plant level safety-related topics include defining multi-plant safety standards, defining safety cooperation levels, defining acceptable multi-plant risks, joint workforce planning in the event of multi-plant accidents, joint emergency planning, etc. In the chemical industry, at factory level and at plant level, safety documents, guidelines, and instructions, technical as well as nontechnical or managerial, are usually well elaborated. However, despite the increasing potential accidental as well as intentional danger represented by many chemical industrial parks worldwide where ever more hazardous chemicals are treated, multi-plant- or cluster-specific safety guidelines do not exist nor do collaboration-intensive strategic multi-plant safety investment strategies exist. This book first discusses the latest insights into approaches to truly advance multi-plant safety management (Chap. 2). Second, our study uses game theory to discuss how decision makers from different chemical plants may be encouraged to engage in such collaborative initiatives (Chaps. 3–7).

1.2.2 Multi-Plant Safety Background: The Existence of **External Domino Effects**

A wide variety of chemical risks exists. The most dangerous ones (situated in the very high severity and very low probability range) are called domino risks, a term by which the potential for a knock-on interaction between groups of installations in the event of an accident at one of the installations is connoted. This mechanism is referred to as 'domino', 'escalation', 'interaction', or 'knock-on'. Domino risks or the risks associated with domino effects (i.e., domino accidents), have a very high destruction potential. The study of domino effects is performed by investigating the different successive accidents, so-called domino events, which constitute a domino effect (Lees 1996; Delvosalle 1996; Council Directive 96/82/EC 1997).

While they have been recognized for a long time, the literature remains relatively vague on the domino effect subject. There is still no generally accepted definition of what constitutes domino effects, although various authors have provided suggestions. Table 1.1 presents an overview of the current definitions identified in a review of relevant documents.

It should be noted that some definitions from Table 1.1 cannot be applied to all conceivable escalation events, but that these definitions are restricted to a specific

Table 1.1 Nonexhaustive list of domino effect definitions

Author(s)	Domino effect definition
Third report of the advisory committee on major hazards (HSE 1984)	The effects of major accidents on other plants on the site or nearby sites
Bagster and Pitblado (1991)	A loss of containment of a plant item which results from a major incident on a nearby plant unit
Lees (1996)	An event at one unit that causes a further event at another unit.
Khan and Abbasi (1998)	A chain of accidents or situations when a fire/explosion/missile/toxic load generated by an accident in one unit in an industry causes secondary and higher order accidents in other units
Delvosalle (1998)	A cascade of accidents (domino events) in which the consequences of a previous accident are increased by the following one(s), spatially as well as temporally, leading to a major accident
AIChE-CCPS (2000)	An accident which starts in one item and may affect nearby items by thermal, blast or fragment impact
Vallee et al. (2002)	An accidental phenomenon affecting one or more installations in an establishment which can cause an accidental phenomenon in an adjacent establishment, leading to a general increase in consequences
Council Directive 2003/105/EC (2003)	A loss of containment in a Seveso installation which is the result (directly and indirectly) from a loss of containment at a nearby Seveso installation. The two events should happen simultaneously or in very fast subsequent order, and the domino hazards should be larger than those of the initial event
Post et al. (2003)	A major accident in a so-called 'exposed company' as a result of a major accident in a so-called 'causing company'. A domino effect is a subsequent event happening as a consequence of a domino accident
Cozzani et al. (2006)	Accidental sequences having at least three common features: (1) a primary accidental scenario, which initiates the domino accidental sequence; (2) the propagation of the primary event, due to "an escalation vactor" generated by the physical effects of the primary scenario, that results in the damage of at least one secondary equipment item; and (3) one ore more secondary events (i.e., fire, explosion and toxic dispersion), involving the damaged equipment items (the number of secondary events is usually the same of the damaged plant items)
Bozzolan and Messias de Oliveira Neto (2007)	An accident in which a primary event occurring in primary equipment propagates to nearby equipment, triggering one or more secondary events with severe consequences for industrial plants

(continued)

Table 1.1 (continued)

Author(s)	Domino effect definition
Gorrens et al. (2009)	A major accident in a so-called secondary installation which is caused by failure of a so-called external hazards source
Antonioni et al. (2009)	The propagation of a primary accidental event to nearby units, causing their damage and further "secondary" accidental events resulting in an overall scenario more severe than the primary event that triggered the escalation

Source Reniers (2010b)

type of domino effects, that is, domino effects involving more than one plant. Different types of domino effects can thus be distinguished. Whereas 'internal domino effects' denote an escalation accident happening inside the boundaries of one chemical plant, 'external domino effects' indicate one or more knock-on events happening outside the boundaries of the plant where the domino effect originates, as a direct or as an indirect result.

Although the consequences of external domino effects can be devastating, this phenomenon has so far attracted very little attention of prevention managers in existing chemical clusters. The reason for this rather strange observation is two-fold. On the one hand, modeling of domino effects is highly complex, that is, this type of event is characterized with highly complex cause–consequence relationships which cannot be modeled, nonlinearities, and tight coupling. On the other hand, the probability of domino accident events is extremely low. In order to assess domino effect consequences, deterministic models have to be used in combination with probabilistic models. The main problem about deterministic modeling arises from the transient nature of the events. The difficulties in the application of probabilistic models are in particular due to the fact that the original input data for the probabilistic analyses are often missing. There is simply not enough accurate information available to use 'standard' models which are used for accidents showing higher frequencies. It is thus obvious that such multi-plant domino effects make for one of the most complex accident types existing to date to study and to tackle within an industrial setting.

Nonetheless, in this book, external domino effects are particularly of interest because such accidents involve different companies situated within a chemical cluster, and strategic (prevention investment) choices to avoid—or not—these types of accidents, need to be made by different plants. Simple approaches to risk management and decision making that may work well in a single-plant context, cannot be implemented in a multi-plant context due to a number of possible reasons contained in strategic decision making (e.g. lack of trust between players, confidentiality concerns between players, players' belief that joint prevention investments will be very high, players' belief that collaboration will lead to increased costs, etc.). Therefore, an effective approach is needed to convince the companies to strategically collaborate with each other as regards multi-plant safety prevention and to persuade them to join forces in taking investment decisions in this complex risk area.

1.3 Safety Versus Security: A Brief Description

In order to have a very clear understanding of the topic this book is dealing with, safety, the differences between safety and security are briefly explained in this section. Safety and security are two related concepts but they have a different basis. Table 1.2 gives an overview of various definitions for safety and security (Reniers et al. 2011). A distinction is made between definitions that focus on specific properties and definitions that focus on global properties.

Safety and security are thus different in the nature of incidents: safety is non-intentional, whereas security is intentional (and related with deliberate acts). This implies that in the case of security an aggressor is present who is influenced by the physical environment and by personal factors. These parameters should thus be taken into account during security assessments. The aggressor may act from within the organization and/or from outside the organization. Probabilities in terms of security are very hard to determine. Hence, the identification of threats and the development of measures in terms of security is a challenging task, differing from taking prevention measures for safety reasons.

Both concepts also differ in their approach. In case of safety assessments (or so-called 'risk analyses'), risks are detected and analyzed by using consequences and probabilities (or frequencies). In case of security assessments (or so-called 'threat assessments'), threats are detected and analyzed by using consequences, vulnerabilities, and target attractiveness. The different approach sometimes leads to the need for different and complementary protection measures in case of safety and security. Table 1.3 provides an overview of the different characteristics attached to safety and to security (Reniers et al. 2011).

Domino risks can thus either be safety risks, implying having an accidental nature, or security risks, implying having an intentional nature. Hence, (internal and external) domino effects either happen by chance, or they can be deliberately induced. Despite the fact that all domino accidents that have happened thus far are believed to have been accidental, companies forming a chemical cluster such as the very large chemical clusters of Houston (USA), Antwerp (Belgium),

Table 1.2 Definitions of safety and security from a specific and a global viewpoint

Safety	Protection against human and technical failure
	Harm to people caused by arbitrary or nonintentional events
	Natural disasters, human error or system or process errors
Security	Protect against deliberate acts of people
	Loss caused by intentional acts of people
	Intentional human actions

Table 1.3 Nonexhaustive list of differences between safety and security

Safety	Security
The nature of an incident is an inherent risk	The nature of an incident is caused by a human act
Nonintentional	Intentional
No human aggressor	Human aggressor
Quantitative probabilities and frequencies of safety-related risks are available	Only qualitative (expert-opinion based) likelihood of security-related risks may be available; quantitative probabilities are very hard to establish
Risks are of rational nature	Threats may be of symbolic nature

Rotterdam (The Netherlands), Tarragona (Spain), etc., should be aware of the possibility of someone intentionally causing a domino effect.

Having said this, game theory in this book is employed to investigate the incentives to stimulate strategic cross-plant cooperation investment initiatives for dealing with domino risks related to safety. However, the mathematical technique might also be used to tackle domino risks related to security. To this end, a different model than the one elaborated and explained in this book, would have to be worked out: as mentioned before, protecting installations against intentional attacks is fundamentally different from protecting against random accidents or acts of nature. Intelligent and adaptable adversaries may try different offensive strategies or adapt their tactics in order to bypass or circumvent protective security countermeasures and exploit any remaining weaknesses. The book provides an easy-to-use approach only to deal with safety risks.

1.4 Game Theory Applied in Safety Management: Concise Literature Overview

Nadeau (2003) uses game theory to discuss cooperation in health and safety, demonstrating that incompatibilities exist in the aims and strategies of the social partners with respect to health and safety. Lo et al. (2006) present a game theory-based exit choice model for evacuation. The model has been integrated in an evacuation model and demonstrates that the evacuees' interaction can affect the evacuation pattern and clearance time of a multiexit zone. Chew et al. (2009) show how a game theory-based approach can be used to analyze the interaction of participating companies in an eco-industrial park seeking to develop an inter-plant water-integration scheme. Xiao-Ping et al. (2009) present a systematic framework for industrial symbiosis networks (ISN) evaluation and modeling. The authors propose a mathematical multiobjective game model to enhance the understanding of ISN issues and motivate improvements of economic, resources, and environmental sustainability. Results suggest that the effectiveness of policies designed to improve chemical industrial park management may be enhanced by this proposed framework. These are some examples of game theory used as a mathematical tool in operational safety-related topics. Zhao et al. (2012) use game theory to describe strategy selection for environmental risk and carbon emission reduction. Zhao et al. (2012) also provide an approach to achieve more environmentally friendly products, using game theory to understand the possible actions of government and manufacturers.

Since only Reniers (2010b) has paid some attention to the multi-plant safety (and security) problem, this is thus a research area deserving much more attention from the academia. Organizing robustness within chemical industrial clusters by prevention through collaboration may prove to be extremely important in the future for avoiding huge human and economic losses.

References

AIChE (American Institute of Chemical Engineers) (2000) Guidelines for chemical process quantitative risk analysis, 2nd edn. CCPS (Centre for Chemical Process Safety), New York

Antonioni G, Spadoni G, Cozzani V (2009) Application of domino effect quantitative risk assessment to an extended industrial area. J Loss Prev Process Ind 22(5):439–449

Bagster DF, Pitblado RM (1991) The estimation of domino incident frequencies—an approach. Process Saf Environ Prot 69(B):195–199

Bozzolan J.-C, Messias de Oliveira Neto J (2007) A study on domino effects in nuclear fuel cycle facilities. International Nuclear Atlantic Conference—INAC 2007, Santos, Brazil

Chew IML, Tan RR, Foo DCY, Chiu ASF (2009) Game theory approach to the analysis of inter-plant water integration in an eco-industrial park. J Cleaner Prod 17(18):1611–1619

Child J, Faulkner D, Tallman S (2005) Cooperative strategy: managing alliances, networks and joint ventures, 2nd edn. Oxford University Press, New York

Council Directive 96/82/EC on the control of major-accident hazards involving dangerous substances (1997) Off J Europ Union L010: 13–33

Council Directive 2003/105/EC (2003) Seveso II directive on the control of major-accident hazards involving dangerous substances with amendments. Off J Europ Union L345:97–105

Cozzani V, Gubinelli G, Salzano E (2006) Escalation thresholds in the assessment of domino accidental events. J Hazard Mater 129(1-3):1–21

Curzio AQ (2002) Dynamics and models in theory and practice. Physica-Verlag, Heidelberg

Delvosalle C (1996) Domino effects phenomena: definition, overview and classification. In: Direction Chemical Risks (ed) European seminar on domino effects. Federal Ministry of Employment, Brussels, pp 5–15

Delvosalle C (1998) A methodology for the identification and evaluation of domino effects. Rep. CRC/MT/003, Belgian Ministry of Employment and Labour, Brussels, Belgium

Gorrens B, De Clerck W, De Jongh K, Aerts M (2009) Domino effecten van en naar Seveso-inrichtingen. Rep. 07.0007, Flemish Ministry of Environment, nature and energy, Brussels, Belgium

HSE (Health and Safety Commission) (1984) The control of major hazards, third report of the HSC advisory committee on major hazards, HMSO, London

Khan FI, Abbasi SA (1998) Models for domino effect analysis in chemical process industries. Process Saf Prog 17(2):107–113

Lees FP (1996) Loss prevention in the process industries, 2nd edn. Butterworth-Heinemann, Oxford

Lo SM, Huang HC, Wang P, Yuen KK (2006) A game theory based exit selection model for evacuation. Fire Saf J 41(5):364–369
Nadeau S (2003) Cooperation in health and safety: a game theory analysis. Law Rev (Risks) 1(3/4):219–227
Post JG, Bottelberghs PH, Vijgen LJ, Matthijsen AJCM (2003) Instrument Domino Effecten. RIVM, Bilthoven
Reniers GLL (2010a) Multi-plant safety and security management in the chemical and process industries. Wiley-VCH, Weinheim
Reniers GLL (2010b) An external domino effects investment approach to improve cross-plant safety within chemical clusters. J Hazard Mater 177:167–174
Reniers GLL, Cremer K, Buytaert J (2011) Continuously and simultaneously optimizing an organisation's safety and security culture and climate: the improvement diamond for excellence achievement and leadership in safety and security (IDEAL S&S). J Cleaner Prod 19(11):1239–1249
Straffin PD (1993) Game theory and strategy. The Mathematical Association of America, Washington DC
Xiao-Ping J, Fang W, Shu-Guang X (2009) Multi-objective game models for chemical industrial park. Comput Aided Chem Eng 27:2019–2024
Vallee A, Bernuchon E, Hourtolou D (2002) MICADO: Méthode pour l'identification et la caractérisation des effets dominos. Rep. INERIS-DRA-2002-25472, Direction des Risques Accidentels, Paris, France
Zhao R, Neighbour G, Han J, McGuire M, Deutz P (2012) Using game theory to describe strategy selection for environmental risk and carbon emissions reduction in the green supply chain. J Loss Prev Process Ind 25(6):927–936

Chapter 2
Safety Management in Chemical Industrial Clusters: The State of the Art

2.1 Different Types of Accidents Within a Chemical Industrial Area

Some background information on different types of accidents and various safety managerial practices within chemical industrial plants and parks is required before the need for game theory in the chemical industry can be well-understood and before recommendations on how to use the elaborated and explained models in this book, can be understood and applied in industrial practice.

Three kinds of accidents can be distinguished: accidents where a lot of historical data is available (we call them type I), accidents where little or extremely little historical data is available (we call them type II), and accidents where no historical data is available (we call them type III). Consequences of type I accidents mainly relate to individual employees (e.g., most work-related accidents), the outcome of type II accidents may affect a company or large parts thereof (e.g., large explosions, internal domino effects, and some external domino effects), and type III accidents for example have an impact upon several chemical plants at once within an industrial park (e.g., external domino effects with huge consequences or terrorist-induced external domino effects).

Whereas type I accidents imply most work-related accidents such as falling, little fires, slipping, etc., type II accidents can be regarded as catastrophes with major consequences and often with multiple fatalities. Type II accidents do occur on a fairly regular basis in a worldwide perspective, and large fires, large releases, explosions, toxic clouds, etc. belong to this class of accidents. Type III accidents are 'true disasters' in terms of the loss of lives as well as of economic devastation. These accidents often become part of the collective memory of humankind. Examples include disasters such as Seveso (Italy, 1976), Bhopal (India, 1984), Chernobyl (USSR, 1986), Piper Alpha (North Sea, 1988), 9/11 terrorist attacks (USA, 2001), and more recently Deepwater Horizon (Gulf of Mexico, 2010), and Fukuchima (Japan, 2011). As may be noticed by the reader, besides an important unwanted internal outcome, such type of accidents also often have huge consequences outside the organization where the accident originated.

G. Reniers and Y. Pavlova, *Using Game Theory to Improve Safety within Chemical Industrial Parks*, Springer Series in Reliability Engineering,
DOI: 10.1007/978-1-4471-5052-7_2, © Springer-Verlag London 2013

For preventing type I accidents, risk management techniques and practices are widely available. Statistical analyses based on past accidents can be carried out to predict possible future type I accidents, indicating the prevention measures that need to be taken to prevent such accidents.

Type II accidents are much more difficult to predict, and thus to avoid. They cannot be predicted via common statistical analysis since the frequency with which these events happen, is too low. Nevertheless, some information about how to deal with this type of accidents is available, but widely used statistical methods cannot be employed to investigate this information. The errors of probability estimates are simply too large to be able to use such probabilities (Taleb 2007). Hence, managing such risks is based on the scarce data that are available and on extrapolations, assumptions, and expert opinions.

The third type of accidents are simply impossible to predict. No information is available about them and they only happen extremely rarely, as a result of a series of simultaneously occurring circumstances which themselves individually are rare. Such accidents can also be called 'black swan accidents'.

Type I accidents from Fig. 2.1 are non-major accidents (for example accidents resulting in the inability to work for several days, accidents requiring first aid, etc.) and cannot be related to external domino accidents. Therefore, risk management collaboration between companies is not strictly necessary to manage such accidents and these accidents are not further discussed in this book. The reader should, however, realize that preventing such accidents will also be more efficient through—although not required—cross-plant collaboration and the accompanying exchange of best practices and risk management know-how and knowledge.

Type II accidents and type III accidents can both be categorized as major accidents (e.g., multiple fatality accidents, accidents with huge economic losses, etc.).

Based on the different types of risks, and on new insights due to major accidents such as the BP Texas City refinery disaster of 2005, the well-known 'accident pyramid' needs to be improved. Heinrich (1950), Bird (Bird and Germain 1985), and Pearson (James and Fullman 1994) all established a pyramidal relationship between the number of fatalities, the number of permanently injured, and the number of nonpermanently injured (first aid cases etc.). Figure 2.2 displays such an accident pyramid form.

The ratios found by the different researchers were different (varying from 1:300 to 1:600) depending on the industrial sector, the research area, cultural aspects,

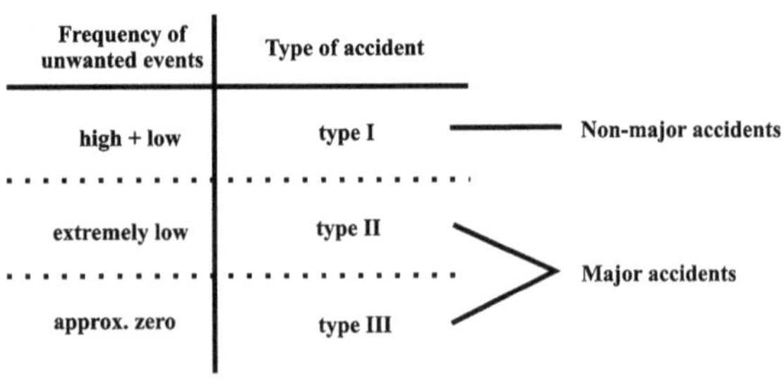

Fig. 2.1 Frequency of unwanted events in relation to the different types of accidents

Fig. 2.2 Bird pyramid

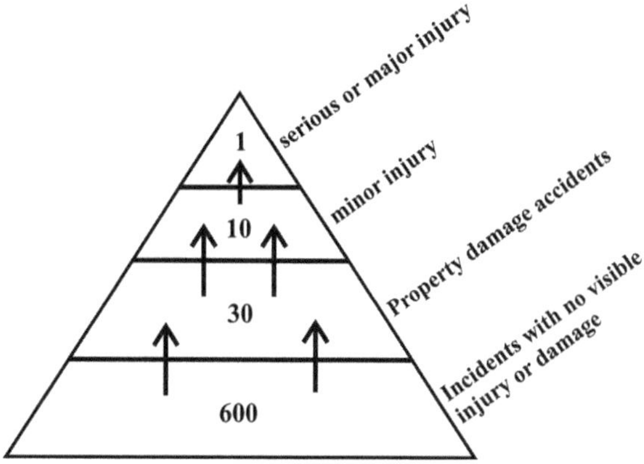

etc., but qualitatively the accident pyramid was reproduced by empirical evidence and thus exists.

However, as mentioned before, improvement of the form of the pyramid is needed. Instead of an 'Egyptian pyramid', the structure of the pyramid should rather be more shaped as a 'Mayan pyramid', showing a stair tread shape as shown in Fig. 2.3.

The Mayan pyramid indicates the difference between unwanted events leading to occupational accidents, and unwanted events leading to disasters and catastrophes. Not all near-misses have the potential to lead to disaster, only a small fraction of unwanted events may factually turn into a catastrophe. Hence, if one desires to prevent disasters (of type II and III), safety management should be aimed at these types of unwanted events as well, and not merely at the large majority of unwanted events possibly leading to 'common' accidents. Hopkins (2010) uses a 'two pyramid' model to indicate the latter, one pyramid representing

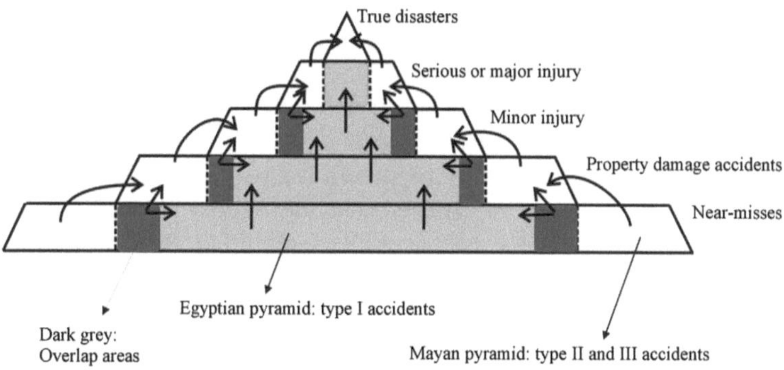

Fig. 2.3 Mayan pyramid shape (improved version of accident pyramid; from T. Meyer, G. Reniers, Engineering Risk Management, 2013, with kind permission © De Gruyter GmbH)

'common' accidents, and a second pyramid next to it (and partially overlapping) standing for accidents which may lead to true disaster. The pyramids overlap partially at the bottom, since some events may be warnings and incidents for all types of accidents.

Despite all risk management practices available to date, type II and type III accidents can be expected to happen in the future, even with higher frequencies and with more devastation and losses of lives than has ever been the case in the past. The reasons are simple and can easily be understood:

1. current risk management practices are aimed at different types of risks and are inadequate and insufficient for type II or type III accident risks;
2. chemical industrial areas are ever more integrated and complex, characterized with complex cause-consequence relationships that cannot be modeled by any currently available accident model;
3. the number of chemical substances used (that is, produced, stored, processed, and/or transported) within chemical industrial areas increases year-by-year (throughout the world);
4. the number of people working within, or in the vicinity of, such chemical industrial areas rises year-by-year (throughout the world).

The first reason results from the fact that domino effects can be categorized as either type II or type III accidents. This indicates that traditional risk analysis and risk management techniques cannot be employed to deal with this type of risks, simply because not enough information is available. Hence, to deal with this type of risks, a new kind of management should be installed in chemical industrial areas: collaborative multi-plant safety management. The next sections provide more information about this kind of innovative safety management.

The second reason follows from insights in what is called Normal Accident Theory (NAT). NAT predicts 'system accidents', arguing that if the organization is 'complexly interactive' rather than linear, and 'tightly coupled' rather than loosely coupled, small errors can interact in unexpected ways and the tight coupling will mean a cascade of increasingly large failures. A failure in a linear system (e.g., an assembly line) is anticipated, comprehensible, and generally visible. It can be fixed and if the precaution measures are carried out properly, the accident probably will not happen again in the linear system. Complexly interactive systems can have independent failures, each insignificant in itself, that interact in unexpected and even incomprehensible ways such as to evade or defeat the safety devices set up to respond to the individual failures. If the system is also 'tightly coupled', the initial failures cannot be contained or isolated and the system stopped; failures will cascade until a major part of the system or all of it will fail. Readers who might be interested in NAT are referred to Perrow (1999, 2006).

The third and fourth reasons simply indicate that the consequences may be much more severe in the present than in the past. More hazardous substances may lead to a more disastrous outcome, and more people present may lead to more fatalities, if a major accident would occur.

2.2 Safety Management in Chemical Clusters: State-of-the-Art Research and Insights

The emphasis of enhancing safety in a chemical multi-plant context (in order to avoid cluster-related accidents or mitigating the consequences of such accidents) lays on the prevention of external domino effects. To identify which chemical plants pose an escalation risk or threat to one another and to which extent software and instruments are internationally being developed, the interested reader is referred to Reniers et al. (2006). The tools are either used to identify the potential for domino effects and/or the resulting scenarios of major accidents. However, these tools are used for taking measures to prevent type II accidents (e.g., most internal domino effects). It would not be useful or possible to use such techniques on type III accidents, since once such an accident has happened, information about the accident becomes available, and the subsequent similar accident becomes a type II accident (due to the nature of the definition of type III accident).

Currently, many companies assume that preventing internal domino effects suffices to prevent external domino effects, the reason being that the probability of an initiating internal domino effect is several times higher than the probability of a subsequent external domino effect (and hence, the latter domino effect is simply disregarded by most risk experts as 'too improbable'). It is obvious that if the probabilities of higher-order domino effects are taken into account or if the installations within the entire chemical industrial area are all taken into account in one go during the probability exercise, this current reasoning should be abandoned. Moreover, type III accidents are not yet known by risk consultants or company risk managers, and external domino effects are insufficiently recognized by risk experts to have the potential to become a true type III accident.

If risk consultants or company risk experts would recognize the true existence of extremely rare events, they would realize that current risk management practices are insufficient to deal with them and that other risk management ways are needed. If one plays Russian roulette with a revolver containing 100,000,000 chambers with only 1 bullet, after playing every minute for a number of years one forgets that he/she is playing Russian roulette, taking even larger risks (adding some bullets to the revolver, e.g., by increasing the complexity of operations), and at some point in time (it may be tomorrow, it may be in ten years) a bullet will be fired, 'completely unexpected'. Without reason, the accident will then be considered as an unpredictable and thus unavoidable side-effect of playing roulette, or, if translated into risk management of chemical clusters, of chemical production. However, type III accidents in the form of external domino effects can be dealt with! The required and most important approach to tackle such accidents, is first and foremost the recognition of their existence, and subsequently safety collaboration. Collaboration between companies to prevent external domino effects from happening would indeed be much more efficient and effective (cost-effective as well as prevention-effective) than the current (insufficient) one-company way of dealing with the prevention of potential cluster-ruining accidents.

How can such collaboration be organized and be elaborated in a workable and efficient way? Reniers (2010) suggests to draft a safety management system on a multi-plant level. Of course, there is the question as to who will perform the risk assessment study, how will it be performed, and how will the implications of the study and the proposed preventive measures be executed. Personnel with specific expertise in terms of the cluster, but independent to the cluster plants, would be able to handle confidential information and to provide suggestions.

To structure safety issues at a multi-plant level, multi-plant-related coordination is to be organized by a 'Multi-Plant Council' (abbreviated 'MPC'), grouping organization representatives from participating plants and independent delegates. Hence, the Multi-Plant Council should be built-up as illustrated in Fig. 2.4.

The MPC thus consists of plant safety representatives on the one hand, and of independent experts on the other hand. The single plant expert participants have a typical counseling function, formulating safety recommendations as a result of joint think tank brainstorming and communication sessions. The independent experts need to gather confidential information, to be able to use this info for example while implementing the game-theoretic models elaborated in this book. The MPC Data Administration is composed of independent consultants (i.e., impartial knowledgeable personnel) responsible for administering all necessary (open as well as confidential) safety-related information gathered from the different plants of the cluster.

As explained by Reniers (2010), brainstorming sessions for the plant departments of e.g., Production, Maintenance, Human Recourses, Logistics, Environment, Management, Safety Management, Security Management, etc. may be organized in so-called "MPC think tanks". The number of participants of such a think tank should deliberately be limited with a view to maximizing output efficiency. If the number of participating plants in the multi-plant initiative is too

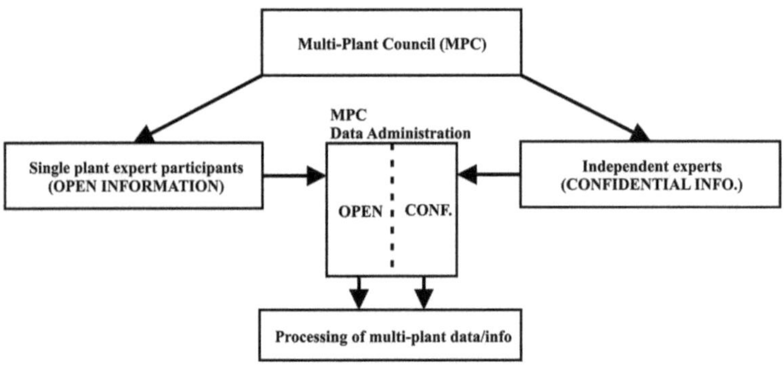

Fig. 2.4 Multi-Plant Council architecture

large, a method of systematic alternation of representatives can be used. As a guideline, maximum eight representatives per think tank (except for the safety and security think tank) is suggested by Reniers (2010): seven department representatives (after a fixed period of time one representative after another alternating with department representatives from the other companies) and one independent consultant with expertise in the department field. The MPC safety management think tank should, according to Reniers (2010), aim at achieving integrated preventive multi-plant safety, by drawing and proposing standardized procedures (together called the Multi-Plant Safety Management System or M-PSMS) based on plant department recommendations and added multi-plant safety issues. The crucial role of the safety management think tank is reflected in its composition, i.e., permanent multi-plant safety specialists from the most dangerous companies are added to the small group of eight. The safety management think tanks maximum consists of 12 group members, the numbers again being limited for the same reason as given previously. The M-PSMS is translated and implemented at individual plant level by the departments. Hence, a continuous improvement of drafting the different parts of the M-PSMS and the single-plant SMSs is achieved by optimized communication and cooperation of every department of the plants participating to the multi-plant initiative.

The collection of multi-plant data (e.g., incidents, accidents, etc.), e.g., employed for making multi-plant recommendations by employing game-theoretic models, can be supported by the MPC Data Administration. Such an administration works closely together with the safety management think tank and includes independent external safety-, security-, and game theory experts.

By splitting the Multi-Plant Council into a part composed of single plant experts, i.e., the think tanks, and a part composed of independent experts, balance between confidentiality and information sharing is targeted. The think tanks and the multi-plant organizational platform ensure the continuous improvement in taking preventive measures as regards multi-plant topics and, to a lesser extent, single plant safety topics. The independent experts collect the necessary confidential financial information, and all other sensitive information and use all this information for example as input for executing the game-theoretic models further explained in this book. Based on the output, the MPC gives guidance and recommendations to the strategic decision makers deciding about domino effects prevention investments. If calamities should occur having a possible impact beyond the originating company, the necessary data of all the plants is centralized in an MPC Data Administration databank and can be used without any delay.

More information on the different approaches and frameworks required to install such an MPC and to make it workable in real industrial practice, is found in Reniers (2010). The same author also provides straightforward and easy-to-implement recommendations for achieving multi-plant safety and security loops of continuous improvement embedded in a cross-plant safety and security management system.

Empirical research actually shows that chemical companies recognize the necessity for improved cooperation (Reniers et al. 2005). Companies are convinced of the safety maximizing synergy effects of cross-company safety risk analysis, but at the same time openly question the feasibility of more intensive cooperation for five main reasons. First, companies belonging to an international group with standard safety methods are often obliged to use these methods. Second, companies with divergent core activities need to be convinced of the safety gains of joint safety management. Several companies participating to the survey conducted by Reniers were, however, convinced that joint training courses and safety drills would improve safety. Third, the desire to collaborate is often limited by practical problems, such as the procedure to purchase personal safety equipment, etc. The fourth reason is the division of the costs of joint prevention measures, especially where mutual risks are not equally divided over the plants and are difficult to measure. The fifth main reason for hesitating to intensively collaborate with nearby plants may be summarized as liability questions. These considerations and the confidentiality of company safety data are the major hurdles facing collective safety risk analysis in the chemical sector.

Cooperation between distrustful rival businesses can be enhanced if they are faced with a common enemy (e.g., terrorism). Security issues have the advantage of fulfilling the latter requirement. However, security management also works to prevent highly confidential company information from becoming known to competitors. Moreover, costs made to assure joint plant security also seem to be a major practical problem to overcome. Therefore, although chemical businesses face new security threats and respond to them with relatively new and fairly cross-enterprise standardized security vulnerability analysis, improved interfirm security collaboration is subject to the same obstacles as those mentioned for enhancing safety collaboration. But, as already indicated, this book will focus on game theory for enhancing multi-plant collaboration as regards safety matters, and does not treat security collaboration issues.

If companies' top-management (influenced and informed by risk consultants and company risk management) would truly recognize the hypothetical benefits of avoiding type II and III accidents as well as their potential companies' ruining consequences, cross-plant collaboration would surely be more practised. Nonetheless, there are some signs that consciousness about multi-plant cooperation is rising and improvements are made. Current industrial practice shows that informal gatherings of companies belonging to the same industrial cluster are (slowly) evolving toward more formal organizations discussing operational practices. Examples are the 'Deltalinqs University' which was founded in the port of Rotterdam in 2001 and the 'Delta Process Academy' (DPA) founded in the port of Antwerp in 2005, ports hosting two major industrial chemical clusters in Europe. Although these initiatives are still quite informal, they express the willingness of the major players to cooperate more intensively and on a larger scale in the field of operational safety. However, multi-plant safety and security cooperation also requires intensive collaboration at strategic level (supported by top level management) where it should be guaranteed that specific data is treated confidential. These objectives can be achieved by the suggested Multi-Plant Council. Using existing best industrial practices from single chemical companies and integrating them into a best cross-industrial practice for continuously improving safety and security within industrial areas is believed by experts to be the only possible methodology/way for bringing theory into practice. The suggested approach indeed might lead to a safety situation of a chemical industrial park of the next generation.

Commitment and communication are key factors in enhancing cooperation and information exchange to improve internal and external safety. Once multi-plant safety commitment and multi-plant safety communication are accepted as company key notions, multi-plant safety cooperation can be streamlined. The next section discusses the usefulness of game theory in this regard and the need for a mathematical approach in the multi-plant decision-making process on a strategic (investment) level.

2.3 Safety Management in Chemical Clusters: Shaping the Future by Using Game Theory

Persuading management from different neighboring chemical plants to invest in external domino effect prevention and starting up cross-plant safety collaboration on a strategic level, is essential to develop a successful multi-plant initiative. Establishing an effective Multi-Plant Council can only truly be achieved if top-management from various chemical plants fully engage in multi-plant safety on a strategic level. However, no clear and straightforward economic incentives exist to urge companies within chemical clusters to jointly develop multi-plant external domino risk management. To this end, this Section indicates how a broad-based game-theoretical perspective may help to make the financial benefits of external domino effects prevention measures obvious to top-management belonging to different chemical plants.

As already mentioned, risks as regards external domino effects between two chemical plants are risks whose consequences depend on a company's own risk management strategy and on that of the adjacent company. Game theory can be used to predict the external domino effect prevention outcome of a situation where companies make independent decisions on whether to invest in cross-plant prevention or not, but are at the same time aware of the strategic external domino effect decisions (to 'invest' or to 'not invest') made by other companies. Hence, the external domino effect prevention investment choices made by every individual chemical facility might lead to socioeconomic optimal or suboptimal situations. In the context of this book, a socioeconomic optimal situation represents a situation where society and economy are protected as good as possible at the optimal (minimized) cost against the devastating consequences of an external domino effect. If the costs are sufficiently low (so that each company wants to invest in external domino effects prevention, even if the neighboring company did not incur these costs) is a straightforward example of a socioeconomic optimal situation. If external domino effects prevention investments would appear to be very high to both companies relative to their potential benefits, then it might be efficient for no company to incur investment costs (inducing a socioeconomic suboptimal situation).

If a company decides to invest in multi-plant safety, not only the company itself, but also its neighboring company is less vulnerable to an incidental domino effect triggering an external domino effect. Expectations and perceptions about the neighbors' decisions will thus influence investments in multi-plant safety prevention measures. As a result of perception, the socioeconomic outcome might be suboptimal for both companies. This situation of decision making of two neighboring plants can be modeled as what is called a 'game' and—by solving the game—give conditions for a win-win situation where both companies win by investing in multi-plant safety prevention measures. Chapter 3 discusses the essentials of game theory needed for the further understanding of this book. The financial background of the choice 'to invest or not' is illustrated in Fig. 2.5.

The expected investments versus performance curve depicted in Fig. 2.5 qualitatively depicts the investments for multi-plant safety precautions versus the performance and effectiveness of these precautions (representing the safety efficiency of the company). The curve should be upward sloping because the higher the investments are, the more or better safety prevention within the multi-plant area can be afforded, and thus the higher the prevention performance is. Points situated on the curve indicate an efficient planning of precautions: no increase in prevention measures' effectiveness can be obtained without a corresponding increase in investments, and no decrease in investments can be obtained without a decrease in precaution performance. However, points on the investments versus performance curve do not necessarily represent optimal planning. Point A in Fig. 2.5 illustrates the latter: although multi-plant safety precautions are planned

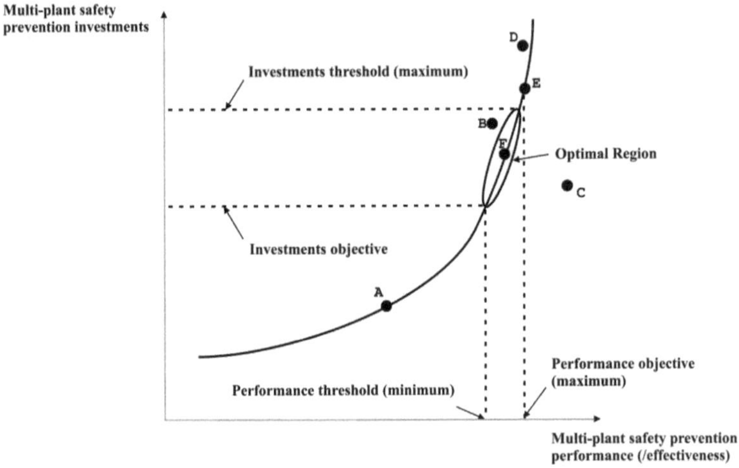

Fig. 2.5 Multi-plant safety prevention measures planning in case of two neighboring chemical companies

very efficiently, their performance is not adequate (below a minimum performance threshold), thus planning is not optimal. Points lying above (or to the left of) the curve indicate feasible, but inefficient, planning. This is indicated by point B in Fig. 2.5. Here, the planning is an inefficient combination of investments and performance, because multi-plant safety prevention could be developed at the same investment cost but with higher performance or at the same level of performance with lower investments (e.g., by cooperating with neighboring companies through using a game-theoretical approach). Points below (or to the right of) the curve (point C in Fig. 2.5) indicate infeasible planning for a given set of input constraints (e.g., functional safety and/or safety limits, technology level limits, safety program limits, etc.).

Most companies are unwilling to invest in multi-plant safety prevention in present industrial settings because they perceive a situation represented by point D in Fig. 2.5 (multi-plant prevention measures taken individually) or point E (multi-plant prevention measures taken by collaboration) in Fig. 2.5. Hence, this would mean a maximum multi-plant safety prevention performance, but at very high expense in excess of the companies' investment costs thresholds. The optimal multi-plant precautions planning should be at an acceptable investment cost with an acceptable performance (point F in Fig. 2.5).

Managing multi-plant prevention investments involve different companies and thus different decision makers. Therefore, using a game-theoretical approach could prove a means of meeting the objective of jointly reaching an acceptable performance at an acceptable expected investment for multi-plant precaution measures.

2.4 Effective Collaboration-Induced Cross-Plant Safety Management

Suppose that game theory would be used and that companies and their top-management are convinced (through the use of game theory) that cross-plant collaboration as regards safety would indeed lead to more effective and more efficient prevention against external domino effects. What could then be essential principles, strategies, and developments of the Multi-Plant Council to manage these extremely low frequency, extremely high consequences risks in an industrial park? In other words, how can be made as sure as possible by the MPC that no type II or type III accident will occur within this industrial cluster?

2.4.1 Multi-Plant Safety Culture and Climate

A company's culture is *a pattern of shared basic assumptions learned by a group as it solves its problems of external adaptation and internal integration, which has worked well enough to be considered valid and, therefore, to be taught to new members as the correct way to perceive, think, and feel in relation to those problems* (Schein 2010). An organizational culture has an impact on the behavior of the employees, the operations and the results of the organization. In order for this impact to be positive, it is important that there is a good fit between the strategy and the culture of the organization (Irani et al. 2004). Besides a company's culture, another important concept is a company's climate. Although both concepts are closely linked, it is imperative to make a clear distinction. Generally speaking, a company's climate can be thought of as the product of some of the underlying assumptions and hence, it is the way in which a company's culture is visible for the outside world. Therefore, a company's climate can be seen as the outer layers of a company's culture and actually the manifestation of the culture. As a result, a company's culture emphasizes continuity, while a company's climate is comparable to a snapshot of the culture of a company.

It is obvious that a multi-plant safety culture and climate needs to be established, where safety topics within the organizational domains of human factors, managerial procedures, and technology are elaborated from a multi-company context, and not merely from a single-plant viewpoint.

Reniers (2010) suggests an integrative safety (and security—but we will further focus on the safety part of his suggestions) culture and climate model consisting of three dimensions. With this model, all safety culture and climate aspects within a multi-plant context can be integrated and covered, since all elements concerning a good safety culture and climate can be placed under one of these three dimensions. The proposed dimensions are People, Procedures, and Technology, and the model is therefore referred to as the P2T-model. The interplay between the three domains defines the present safety culture and climate in any organizational context. A visualization can be found in Fig. 2.6.

One needs little argumentation that in the chemical sector, the technological dimension is indispensable to ensure a good safety culture and climate. With

Fig. 2.6 Safety culture and climate according to the P2T-model, from T. Meyer, G. Reniers, Engineering Risk Management 2013, with kind permission © De Gruyter GmbH

failing installations, a chemical industrial area inflicts a direct threat to all workers present and to surrounding communities. Large improvements in safety technology have been gained in the past decades in many industrial areas. Following the ALARA-principle (As Low as Reasonably Achievable), it should be noted that no risk can be reduced to zero without expenses that can be justified economically. That is why the technological dimension has to be designed in a way that the resulting risk lies between socially accepted boundaries. Governments will impose these bounds, but often chemical enterprises will surpass the required measures.

The second dimension, 'Procedures', is being managed by a (in casu multi-plant) Safety Management System. This management system revises the existing procedures used to maintain a good safety culture. The term 'procedures' can be interpreted very broadly. It concerns procedures to operate safely, to safely store hazardous substances, to manage the competences of employees, and to manage emergency situations, etc. Logically, the organizational structure and culture play a large role in this.

The third dimension to influence safety culture and climate is being defined as 'People'. Reason (1990), Fuller and Vassie (2004), CCPS (2007) and many other researchers indicate that a majority of accidents and near-misses can be attributed to human error. According to some estimates, human error contributes to 90 % of all accidents (Kletz 2001). This number considers all possible sources of error, including front-line operating personnel, engineers, and supervision. That is why creating safety awareness among all industrial area workers is essential for a good safety culture and climate, as well as providing proper training, providing safety incentives, creating a safety-driven multiorganizational community, enhancing competences of employees at all levels, etc.

The P2T model is powerful in its understanding and in its application and can be employed to describe an integrative safety culture and climate. Each of the three dimensions (People, Procedures, and Technology) can be looked upon from the three cultural levels explained by Schein (2010), i.e., artifacts, espoused beliefs and values, and basic underlying assumptions. This way, these three dimensions can be used to measure the safety culture (long-term objectives and results) and safety climate (short-term objectives and results).

A multi-plant area should have a system of continuous safety improvement. One of the most widely used systems in strategic management is the well-known Plan-Do-Check-Act loop of continuous improvement or the so-called Deming cycle. The Deming cycle is a method (originally developed in quality management) that management can use to improve safety within a company. The cycle consists of four steps. The first step (*Plan*) defines the policy that should be followed. Management looks for improvement possibilities and determines how these improvements can be achieved. The planning phase focuses mainly on the long term. The second step (*Do*) implements the elements that were drawn up in the first step. The third step (*Check*) examines if the plans that have been drawn up will be achieved. This phase considers the consequences of the Plan and the Do phases. Finally, the fourth step (*Act*) verifies the results of the plans and examines what can be expected of this plan. These results are then compared to the goals which had been set in the Plan phase and if necessary, adjustments are made and corrective actions are taken. Next, the results of the Act phase can be used to draw up the next plan and the Deming cycle can be used again (Deming 2000). It is only when the whole loop is completed that the continuous improvement system can be considered to be effective.

In current industrial settings, organizations follow the Plan-Do-Check-Act loop of continuous improvement because of their acquired know-how of internationally accepted business standards such as the ISO 9000 series and the ISO 14000 series, addressing quality and environment management systems, respectively. The OHSAS 18000 series, the International Occupational Health and Safety management Assessment System specification that empowers an organization to control its occupational health and safety risks and improve its performance concerning those risks, also uses the PDCA cycle as a basic management concept and is often used to work out safety management systems. The ISO norms and OHSAS are very well-known in the chemical industry and hence, some degree of basic management standardization already exists in this specific industry.

A model to integrate a safety (and security) culture and climate with performance management, leading to company excellence achievement and leadership concerning safety (as well as security) is suggested by Reniers et al. (2011). The proposed model which is called 'Improvement Diamond for Excellence Achievement and Leadership in Safety and Security' (IDEAL S&S), is an adaptation of the P2T-model to visualize the safety (and security) culture and climate in hazardous industries. Figure 2.7 illustrates the model, only focusing on safety.

In order to explain the IDEAL model for safety, a number of terms have to be defined. The IDEAL model shows two fields of tension. The tension between

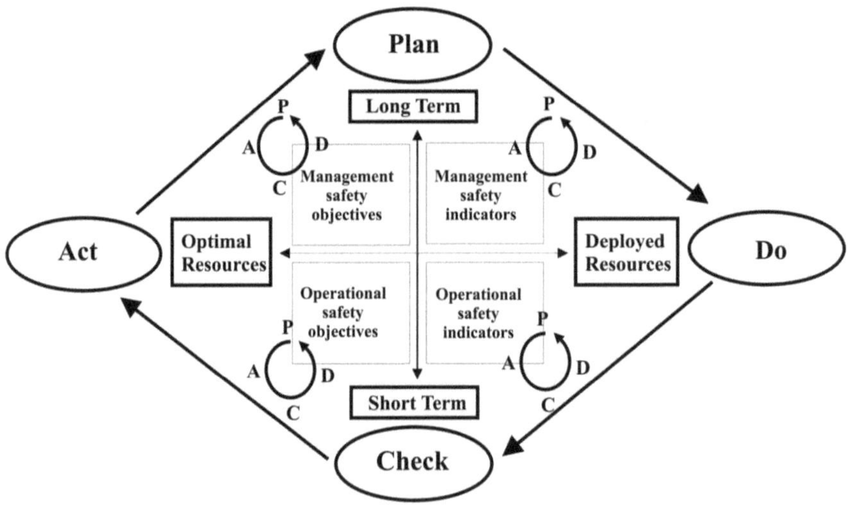

Fig. 2.7 IDEAL model for safety based on Reniers et al. (2011) (with permission from Elsevier, 2013)

Optimal Resources *versus* Deployed Recourses makes up the first field of tension. The term recourses should be interpreted very broadly, e.g., money, knowledge, installations, know-how, people etc. The optimal resources on the one hand are those resources that are necessary to reduce a certain risk component until it reaches the ALARA level. It represents the situation of resources 'as should be in ideal circumstances'. On the other hand, the deployed resources are the recourses which are/can be deployed in reality and thus in the real industrial setting of the plant. It is the situation 'as is in real circumstances'. In an ideal equilibrated situation, the optimal resources are equal to the deployed resources. On the one hand, if the level of optimal resources lays above the level of the deployed resources, a potential hazardous situation can occur. On the other hand, if the deployed resources surpass the optimal ones, a company has wasted resources since the risk level was already reduced below an acceptable level.

A second field of tension exists between Short Term and Long Term. Influencing a safety culture requires goals in the long term which also need to be translated into manageable short-term goals. Hence, achieving the requirements for a multi-plant area's safety culture is visualized in the (long term) upper part of the model whereas the safety climate (which is a snapshot of the multi-plant area's culture) is represented by the (short term) lower part of the model.

The IDEAL model employs safety indicators. These indicators address a qualitative or quantitative value to different safety climate and culture aspects. The long-term deployed resources are situated between the Plan phase and the Do phase of the Deming cycle and can be steered by using management safety indicators. The short-term deployed resources are situated between the Do-phase and the Check-phase of the Deming cycle and can be steered by using operational safety indicators. Both leading and lagging indicators should be used in both cases. Further information concerning developing leading and lagging (a.o. safety) indicators can be found in e.g., HSE (2006) and Parmenter (2007).

The IDEAL model also uses safety objectives. The amount of optimal resources have to make sure these objectives can be met. Management safety objectives and operational safety objectives, respectively, are quantitative figures or qualitative figures set to be achieved by multi-company management for a specific management safety indicator or a specific operational safety indicator, respectively. Management safety objectives (e.g., with a yearly, two-yearly, or five-yearly frequency) and operational safety objectives (e.g., with a weekly, monthly, or three-monthly frequency) are used to control, remediate, and continuously improve the organization's safety achievements, which ultimately lead to a cluster's excellence in safety.

Management indicators and objectives are used to influence and continuously optimize the multi-plant area's safety culture (long-term approach), whereas operational indicators and objectives lead to constant multiorganizational safety climate assessment and improvement (short-term approach). Indicators and objectives themselves should be continuously planned, implemented, checked, and adapted (if necessary) according to the Deming wheel of improvement.

Furthermore, all management and operational safety objectives and indicators should be worked out by using the three dimensions of the P2T-model, i.e., People, Procedures and Technology. The three dimensions are further categorized into different subdimensions and various safety indicators are linked to each of these subdimensions. For all indicators, a set of minimum multiorganizational specific safety objectives can be established.

Every multi-plant area should define its own safety subdimensions which eventually are used to develop management safety indicators and objectives and operational safety indicators and objectives. For more information on the model, its subdimensions, indicators, etc., the interested reader is referred to Reniers et al. (2011).

Based on social system theory, Wu et al. (2008, 2011) studied the potential correlation between safety leadership, safety climate, and safety performance. The results of the statistical analysis indicated that organizational leaders would do well to develop a strategy by which they improve the safety climates within their organizations, which will then have a positive effect on safety performance. We can therefore assume that employing the suggested IDEAL model for safety on a multi-plant level will lead to truly safer chemical clusters that will expose multiorganizational safety leadership. The basic condition for working out the IDEAL model on a multi-plant level is thorough collaboration between the different companies belonging to the cluster. The way to enhance collaboration between the plants, is the use of game theory, as explained in this book.

2.4.2 Framework for Dealing with All Types of Accidents

It is imperative that multi-plant management realizes the existence of type II and type III accidents and that it strives to continuously improve the cluster's safety standing by taking short-term as well as medium-term, long-term, and extremely long-term measures.

Reality shows us day-by-day that risks exist in every domain or activity, from occupational health risks over process risks to reputational risks. Because risks cannot be eliminated, organisations should manage them in a way trying to minimize the negative outcomes and attempting to optimize the positive outcomes. Such management is called *Risk Management* (RM). If the process of Risk Management is carried out in all domains, it is called *Enterprise Risk Management* (ERM) and has to be logic, systematic, and applicable.

Although RM has been developed over time and within many sectors in order to meet diverse needs, the adoption of consistent processes within a comprehensive framework can help to ensure that risk is managed effectively, efficiently, and coherently across an organization.

A variety of frameworks exist to tackle risks. Most of these frameworks are focused on certain domains. This implies that the frameworks are not applicable in all circumstances, on all levels and for all situations and organizations. The most well-known examples of such frameworks include EFQM, ERM, The Balanced Scorecard, lean management, etc. To take all aspects of risks into account and to take optimal decisions, risks should be viewed from a holistic viewpoint, meaning that all stakeholders should be involved (with their own specialism) in the risk management process, and that a dynamism should be brought into the risk decision and expertise process which transforms chaos and ignorance into order and knowledge.

The 'classic viewpoint' of risk management is one of narrow thinking of purely organizational risk management (in the best case ERM), and taking different domains within the company into account, integrating them to take decisions for maximizing the positive side of risks and minimizing the negative side of risks. The process for taking decisions in organizations should, however, be much more holistic than this. Risks are characterized by uncertainties. Hence, all these uncertainties and their possible outcome(s) should be anticipated, that is, identified and mapped, for every risk. If this can be achieved, the most optimal decisions (minimizing the negative outcomes and maximizing the positive outcomes of risks) can be taken. The end-goal is to use all the right people and means at the right time, to manage all existing risks in the best possible way, whether the risks are positive or negative, or whether they are known or not. After all, a part of risk management is 'thinking about the unthinkable'.

Therefore, a number of triangles are employed to build and propose a holistic and integrated framework that allows any organization (also any chemical plant or cluster of chemical plants) to optimize its risk decision and expertise process. The framework is called 'ORDER', which is an acronym for 'Optimizing the Risk Decision and Expertise Rad' (see also Reniers 2012).

Following the well-known PDCA loop of continuous improvement from quality management, a four-step plan (the PDRC loop of continuous improvement) is proposed as the basic structure to serve for the ORDER framework. From a holistic viewpoint, risks, and their uncertainties, outcomes and management, should be the concern of organizations-authorities-academia, these three actors within society forming the first triangle 'risk Policy'. This first triangle should be the cornerstone of solid and holistic risk management, creating the right circumstances and helping to induce collaboration between all parties involved. The second triangle 'Decision' consists of information-options-preferences. It is obvious that decision making always requires information, options, and preferences. Without any one of these three, decisions can simply not be made. Each of the blocks of the 'subsequent triangle' of the rad needs to be aware of the three blocks of the 'previous triangle' of the rad: risk information has to be taken from academia, organizations, and authorities, and the same holds for developing options and mapping or composing preferences. The third triangle 'Risk' includes hazards, exposure, and losses. For the different dimensions of the Risk triangle, the Decision triangle has to be considered. The fourth triangle 'organizational Culture' includes people-procedures-technology. Each of these three domains composing an organizational culture has to take the Risk triangle into consideration. On its turn, the organizational Culture triangle serves as a guide for each of the domains within the risk Policy triangle. Figure 2.8 illustrates the rad of risk decision and expertise.

Fig. 2.8 PDRC Rad of continuous improvement

Hence, risk policy guidelines, rules, etc., made or investigated by authorities, organizations, and academia, should ultimately lead to an efficient and effective organizational risk culture, which, in turn, should be used by risk policy makers, as an input for risk policy guidelines, rules, etc., to continuously improve risk decision making, etc. People, procedures, and technology, forming the backbone of this culture, should be continuously optimized through the PDRC Rad displayed in Fig. 2.8.

An effective and efficient organizational culture implies that organizations are open-minded toward collaborating with other organizations and that they are prepared to search for the optimal way to decrease risks, also multi-plant risks, and act accordingly (e.g., in their investment policies).

2.4.2.1 The 'Considering?': Layer of the ORDER Framework

'Considering?'—triangles can be used to have a holistic indication on how to deal with the PDRC building blocks, and what features should be taken into account when addressing the blocks belonging to the rad of the ORDER Framework. For example, it is recommended that academia-organizations-authorities within the risk Policy triangle are all approached by thinking in the long term, thinking circular, and thinking nonlinear. Hence, as far as risks and risk decisions and expertise are concerned, the way to think about people, procedures, and technology by academia, organizations, and authorities should be long-term oriented, circular, and nonlinear besides (on top of) the current short-term, linear, and cause-consequence thinking. Figure 2.9 illustrates the different 'Considering?'-triangles of the ORDER Framework.

For the Decision-making triangle, the 'Considering?' triangle relates to type I, II, and III events. Thus, when exploring, identifying, and mapping information, options, and preferences from the viewpoints of academia, organizations, and

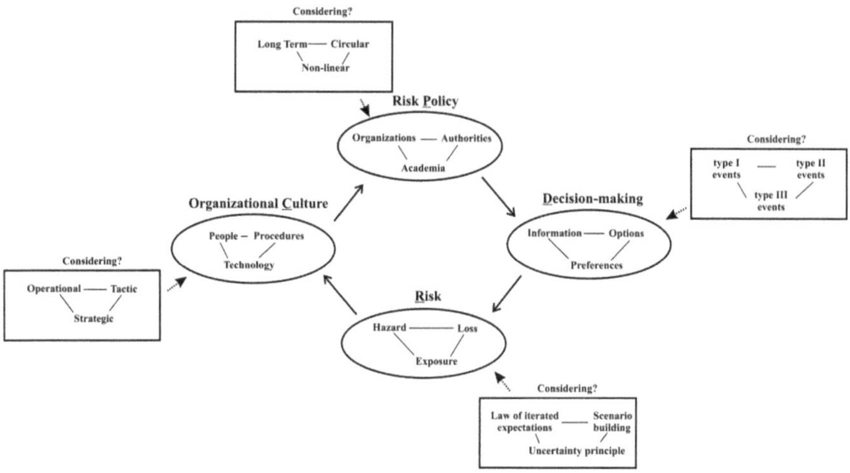

Fig. 2.9 PDRC Rad with 'Considering?'—layer of the ORDER framework

authorities, focus should be on all three types of events. Among others, this implies of course that external domino effects should be taken into account, and information, options, and preferences should be developed regarding such events, even if they are highly improbable.

The way to deal with the Risk triangle for gathering and drafting information, options, and preferences on each of the blocks (that is, possible hazards, possible losses, and possible exposure), should be according to the law of iterated expectations, scenario building, and the uncertainty principle. The law of iterated expectations simply states that if an event (e.g., a domino accident) can be expected somewhere in the future, then the event can also be expected at present. Scenario building is well-known and much used by risk experts. Scenarios are drafted of past events or future events using a variety of available techniques such as risk analysis techniques, the Delphi method, etc. The uncertainty principle (following the precautionary principle reasoning) assumes that is not because a cause-consequence relationship cannot be proven, that it does not exist, in other words, unknown complicated and complex relationships (characterizing any complex phenomenon) should not be disregarded.

People, procedures, and technology (forming the organizational Culture triangle) each need to be considered by an operational, a tactic and a strategic manner of thinking, whereby hazards, losses, exposure, opportunities, and profits should be considered within every domain (people-procedures-technology).

2.4.2.2 The 'Results?': Layer of the ORDER Framework

For each of the triangles of the rad, a 'Results?'—triangle can be drafted. The risk Policy triangle leads to concrete and abstract ideas, rules and guidelines, and laws.

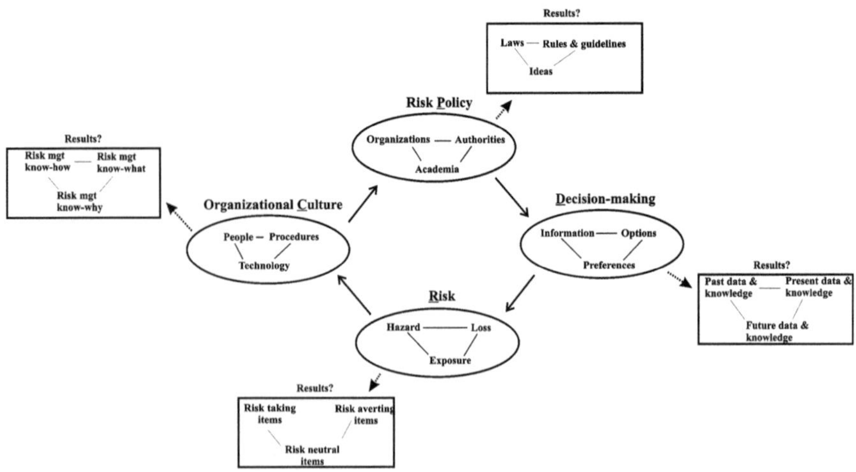

Fig. 2.10 PDRC Rad with 'Results?'—layer of the ORDER framework

The Decision triangle leads to knowledge and information about the past, the present, and the future. The Risk triangle transpires into risk averse-, risk taking-, and risk neutral items of consideration. The organizational Culture triangle leads to risk management know-how, risk management know-what, and risk management know-why in each of its domains. The Rad of continuous improvement is closed by more concrete and abstract ideas, rules and guidelines, and laws. Figure 2.10 illustrates the 'Results?'—Layer of the ORDER Framework.

2.4.2.3 The ORDER Framework

If the triangles from the rad, the 'Considering?'—triangles and the 'Results?'—triangles from the previous subsections are displayed in an integrated way on the same figure, we become a Framework that can be used to continuously advance the optimization of risk decisions and risk expertise within any organization.

Similar to the fact that people do not want physicians only to be able to recognize well-known and 'casual' diseases, and not being able to detect a very rare disease, organizations should not be satisfied with risk experts and risk decisions only tackling well-known and 'usual' (mostly occupational) risks, and not considering out-of-the-ordinary-thinking risks (such as e.g., extremely low frequency, extremely high consequence risks, for example external domino effects) or not using the proper method and the proper data and expertise to tackle certain types of risks. One of the methods to be used for dealing with external domino risks, is game theory.

In managing risks, to elaborate sustainable solutions, it is often much more important (and more difficult) to identify and define in detail the problem(s), than the solution(s). This can only be achieved by using a framework such as displayed

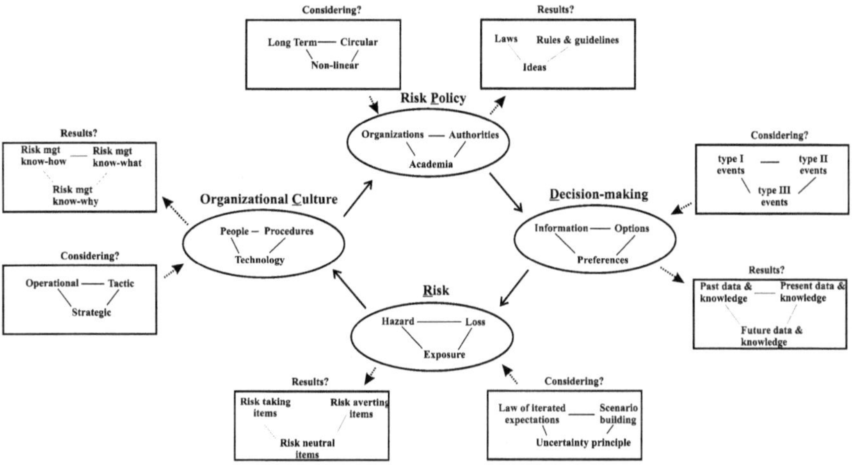

Fig. 2.11 The ORDER framework from T. Meyer, G. Reniers, Engineering Risk Management 2013, with kind permission © De Gruyter GmbH

in Fig. 2.11, integrating all possible viewpoints from diverse stakeholders, risk subjects, methodologies and methods, approaches, disciplines, etc. This book provides an approach to help individual plants to collaborate in an optimal way to solve problems, and thus strives to make risk decisions truly more holistic, more systematic, more objective, and more justified.

Game theory indeed can play an important role to help decision makers and managers of chemical plants to gather and to derive and develop information and options to make better decisions and strategic choices to invest or not to invest in preventive measures. This way, game theory should be regarded as a mathematical tool that is useful with respect to different levels of the ORDER Framework. The next chapter elaborates on game theory as a technique to develop information and options to be more knowledgeable and more informed when taking strategic decisions.

References

Bird E, Germain GL (1985) Practical loss control leadership. The conservation of people, property, process and profits. Institute Publishing, Georgia

CCPS (Center for Chemical Process Safety) (2007). Human Factors methods for improving performance in the process industries. John Wiley and sons, Hoboken, NJ

Deming WE (2000) Out of the crisis. The W. Edwards Deming Institute, Washington, DC

Fuller CW, Vassie LH (2004) Health and safety management: principles and best practice. Prentice Hall, Essex

Heinrich HW (1950) Industrial accident prevention, 3rd edn. Mc GrawHill Book Company, New York

Hopkins A (2010) Failure to learn: the BP Texas City refinery disaster. CCH Australia Ltd, Sydney
HSE Health Safety Executive (2006) Developing process safety indicators. A step-by-step guide for chemical and major hazard industries, Crown
Irani Z, Beskese A, Love P (2004) Total quality management and corporate culture: constructs of organisational excellence. Technovation 24(8):643–650
James B, Fullman P (1994) Construction safety, security and loss prevention. A Wiley Inter science Publication, New York
Kletz TA (2001) An engineer's view of human error, 3rd edn. Institute of Chemical Engineers, Rugby
Meyer T, Reniers G (2013) Engineering Risk Management. In: De Gruyter. Berlin, Germany, book will be published in May 2013
Parmenter D (2007) Key performance indicators: developing, implementing, and using winning KPI's. Wiley, New Jersey
Perrow C (1999) Normal accidents: living with high-tech risk technologies. Princeton University Press, Princeton
Perrow C (2006) The limits of safety: the enhancement of a theory of accidents. In: Smith D, Elliott D (eds) Key readings in crisis management. Systems and structures for prevention and recovery. Routledge, Abingdon
Reason JT (1990) Human error. Cambridge University Press, Cambridge
Reniers GLL, Dullaert W, Soudan K, Ale BJM (2005) Developing an external domino accident prevention framework. J Loss Prev Process Ind 18(3):127–138
Reniers GLL, Ale BJM, Dullaert W, Foubert B (2006) Decision support systems for major accident prevention in the chemical process industry: a developers' survey. J Loss Prev Process Ind 19(6):604–620
Reniers GLL (2010) Multi-plant safety and security management in the chemical and process industries. Wiley-VCH, Weinheim
Reniers GLL, Cremer K, Buytaert J (2011) Continuously and simultaneously optimizing an organization's safety and security culture and climate: the improvement diamond for excellence achievement and leadership in safety and security (IDEAL S&S). J Cleaner Prod 19(11):1239–1249
Reniers GLL (2012) Integrating risk and sustainability: a holistic and integrated framework for optimizing the risk decision and expertise rad (ORDER). Disaster Adv 5(2):25–32
Schein EH (2010) Organizational culture and leadership, VSA. Jossey-Bass, San Francisco
Taleb NN (2007) The black swan: the impact of the highly improbable. Random House, New York
Wu TC, Chen CH, Li CC (2008) A correlation among safety leadership, safety climate and safety performance. J Loss Prev Process Ind 21(3):307–318
Wu TC, Chang SH, Shu CM, Chen CT, Wang CP (2011) Safety leadership and safety performance in petrochemical industries: the mediating role of safety climate. J Loss Prev Process Ind 24(6):716–721

Chapter 3
Introduction into Strategic Decision-Making

3.1 Preliminaries

The purpose of this chapter is to guide readers unfamiliar with the concepts and assumptions of game theory or those wishing to learn more about game theory. We focus on presenting concepts and general results, relevant for analyzing strategic decision-making on cooperative prevention measures against external domino accidents. Game theory can be regarded as an analytical toolbox which helps analyzing the phenomena that can be observed when decision-makers interact. The basic assumptions that underlie the theory are that decision-makers pursue well-defined exogenous objectives (they are 'rational') and that they take into account their knowledge or expectations of other decision-makers' behavior (they reason strategically).

This chapter presents a classification of the games with respect to whether the process is modeled as static or dynamic, with simultaneous moves of players or sequential moves, whether players have complete information, etc., and discuss applicability of these models to the problem of cooperative prevention measures against domino accidents. The material will progress from easy to more difficult, starting from coordination games and the Nash equilibrium concept, and turning to sequential rationality and the concept of 'sub-game perfect solution'. We also discuss players' decision making in conflict situations with incomplete information and Bayesian games.

The contents of the present chapter proceed as follows. Section 3.2 discusses the main features of game-theoretic modeling. It discusses the advantages that this particular mathematical theory brings into the analysis of strategic interaction, but also outlines assumptions specific to game theory and justifies limitations which the theory contains. In Sect. 3.3, we discuss the concept of a 'strategic game' and in Sect. 3.4 the concept of an 'extensive game'. A strategic game is a model of a situation in which each player chooses his plan of action once and for all, and all players' decisions are made simultaneously. The model of an extensive game specifies the possible orders of events; each player can consider his plan of action not only at the beginning of the game but also whenever he has to make a decision.

G. Reniers and Y. Pavlova, *Using Game Theory to Improve Safety within Chemical Industrial Parks*, Springer Series in Reliability Engineering,
DOI: 10.1007/978-1-4471-5052-7_3, © Springer-Verlag London 2013

In Sect. 3.5 we briefly discuss imperfect information games, where the players are moving sequentially but may not be fully informed about each others' moves. Section 3.6 describes games of incomplete information where all or some of the payers have private information, and the section also presents the concept of Bayesian Nash equations.

3.2 Game-Theoretic Modeling

A central feature of multipersonal interaction is the potential for the presence of strategic interdependence. The decision maker faces situations in which their well-being (measured as utility or as profit) depends not only on their own actions but also on the actions of other individuals involved into the situation with strategic interdependence. The actions which are the best for the decision maker may depend on actions which other individuals have already taken, or expected to be taken at the same time, and even on the future actions that other individuals may take or decide not to take, as a result of the current actions.

The tool that we use for analyzing interactions with strategic interdependence is non-cooperative game theory. The term 'game' highlights the theory's central feature: the agents under study are concerned with strategy and winning (in the general sense of expected utility maximization). The agent will have some control over the situation, since their choice of strategy will influence it in some way. However, the outcome of the game is not determined by their choice alone but also depends upon the choices of all other players. This is where conflict and cooperation come into the play. We also emphasize that the term 'non-cooperative game theory' does not mean that non-cooperative theory is unable to explain cooperation within groups of individuals. Rather, it focuses on how cooperation may emerge from rational objectives of the players, given that rationality is a common knowledge, and in the absence of any possibility to make a binding agreement.

It is useful noticing that in the decision-making theory and, consequently, in game theory, individual rationality should not be associated with dispassionate reasoning. We assume that while making an optimal decision, an individual will be able to observe the courses of action that are available and determine the outcomes he/she can receive as a result of the interaction with others. An individual should be able to rank those outcomes in terms of preferences so that the system of preference is internally consistent and can form a basis for choice. Thus, we determine a rational player as one who has consistent preferences concerning outcomes and will attempt to achieve preferred outcomes. Rational play will involve complicated individual decisions about how to choose a strategy which will produce a favorable outcome, knowing that other players are trying to choose strategies which will also produce an outcome favorable to them. It will also involve social decisions about how and with whom a player should try to cooperate.

Game theory is a logical analysis of situations of conflict and cooperation. More specifically, a game is defined to be any situation in which:

- There are at least two players. A player may be an individual, but it may also be a more general entity like a company, an institution, a country, etc.;
- Each player has a number of possible strategies, courses of action which the player may choose to follow;
- The strategies chosen by each player determine the outcome of the game;
- Associated to each possible outcome of the game is a collection of numerical payoffs, one to each player. These payoffs present the value of the outcome to the different players.

We have already mentioned that game theory can generally be divided into non-cooperative game theory and cooperative game theory. Furthermore, we can specify the following classification of games based on the available information and turns when players make their moves. We will look at non-cooperative games which are played only once, involving only a finite number of players, and which give each player only a finite number of actions to choose from. We can distinguish between 'static games' and 'sequential move games'. A static game is one in which a single decision is made by each player, and each player has no knowledge of the decision made by the other players before making their own decision. Sometimes such games are referred to as 'simultaneous decision games' because any actual order in which the decisions are made is irrelevant. A sequential move game is a game where one player chooses his/her action before the others choose theirs. Importantly, the later players must have some information of the first player's choice, otherwise the difference in time would have no strategic effect. Extensive form representations are usually employed for sequential games, since they explicitly illustrate the sequential aspects of a game. Another distinction that can be made between the game-theoretic models is information available to the players. In the models described in Sects. 3.3 and 3.4, the participants are fully informed about each others' moves, while in the models explained in Sect. 3.5 they may be imperfectly informed.

In many economically important situations the game may begin with some player disposing of private information about a relevant issue with respect to his/her decision making. These are called 'games of incomplete information', or 'Bayesian games' (see Sect. 3.6). Incomplete information should not be confused with imperfect information in which players do not perfectly observe the actions of other players. Although any given player does not know the private information of an opponent, he or she will have some beliefs about what the opponent knows, and we will assume that these beliefs are common knowledge. In many cases of interest we will be able to model the informational asymmetry by specifying that each player knows their own payoff function, but that he or she is uncertain about what his or her opponents' payoff functions are.

Traditional applications of game theory attempt to find equilibria in these games. In an equilibrium, each player of the game has adopted a strategy that they are unlikely to change. Many equilibrium concepts have been developed (in particular, the Nash equilibrium) in an attempt to capture this idea. These equilibrium concepts are motivated differently depending on the field of application, although they often overlap or coincide.

We would like to indicate that this methodology is not without criticism, and debates do continue concerning the appropriateness of particular equilibrium concepts, concerning the appropriateness of equilibria altogether, and as regards the usefulness of mathematical models more generally. For instance, game theory should not be considered as a theory that will prescribe the best course of actions in any situation of conflict and cooperation. First of all, the real-world situations are performed in a quite simplified form of a game. In real life it may be hard to say who the players are, to delineate all conceivable strategies and to specify all possible outcomes and it is not easy to assign payoffs. What is typically done is to develop a simple model which incorporates some important features of the real situation. Thus building such a model and its analysis may give insights into the original situation. The second obstacle is that game theory deals with play which is rational. Each player logically analyses the best way to achieve his or her goals, given that the other players are logically analyzing the best way to achieve their goals. In this way 'rational play' assumes rational opponents. It means that players are able to tell which outcomes are more preferred than others and that they are able to align them in some kind of preference relationship order. However, experiments show that in some cases rationality may fail. Game theory results should thus always be interpreted with caution. Finally, game theory does not give a unique prescription for play in games with two players whose interests are not completely opposed (e.g., Stag Hunt game). It also does not have such a prescription for games with more than two players. What game theory offers is a variety of interesting examples, analysis, suggestions and partial prescriptions for these situations. Practitioners using game theory should be aware of this. Hence, by applying game-theoretical concepts and models to enhance safety collaboration in chemical industrial clusters, we obtain (at least partial) insights on how the situation in the cluster as regards collaboration is, and using these insights, recommendations can be formulated.

In provisional summary, a game is a situation where for two or more players their choice of actions has an impact on each other and/or on others. The outcome of a game is expressed in terms of the strategy combinations that are most likely to achieve the players' goals, given the information available to them. Games are often characterized by the way or order in which players move. Games in which players move at the same time or in which the players' moves are hidden from others until the end of the game, are called simultaneous-move games or static games, and they are described in the strategic form. Games in which moves are made in some kind of predetermined order are referred to as sequential-move games or dynamic games, and described in the extensive form. These two types of games will be discussed in the following sections.

3.3 Strategic Form Games with a Discrete Set of Strategies

In this section we study a model of strategic interaction known as a strategic form game (normal form, von Neumann, and Morgenstern 1944). This model specifies for each player a set of possible actions and each decision-maker chooses his/her plan of actions once, and he or she does this for all actions simultaneously, based on a preference ordering over the set of available action profiles. We discuss the Nash equilibrium, the most widely used solution concept for strategic games, in which each player's decision depends on knowledge of the equilibrium.

A strategic game is one in which a single decision is made by each player, and each player has no knowledge of the decision made by the other players before making his/her own decision. Such games are referred to as simultaneous decision games because any actual order in which the decisions are made is irrelevant. To describe the strategic game $\langle N, (S_i), (u_i) \rangle$, we need to specify:

1. the set of players, indexed by $i \in \{1, 2, \ldots, N\}$;
2. a pure strategy set, S_i, for each player;
3. payoffs u_i for each player i for every possible combination of pure strategies used by all players.

Definition 3.1 A **strategy** is a rule for choosing an action at every point at which a decision might have to be made. A pure strategy is one in which there is no randomization.

Usually, no differentiation is made between an action and a strategy in static games.

Definition 3.2 A given preference relation on the set of action profiles of player 1 in a strategic game can be represented by a **payoff** function $u_i : \prod_{i=1}^{N} S_i \rightarrow R$ (also called a utility function), in the sense that $u_i(a) \geq u_i(b)$ whenever a is preferred to b. We refer to values of such a function as 'payoffs' (or utilities).

To keep the notation simple, we will concentrate on two-player games for most of this section. Games with more than two players can be described similarly. In the two-player case, it is conventional to put the strategy of player 1 first and that of player 2 second so that the payoffs to player i are written u_i (s_1, s_2), for any $s_1 \in S_1$ and $s_2 \in S_2$.

When the number of players in the game is small, the strategic form of a game can receive a tabular description. Additionally, it is worth to note that we will focus mainly on finding the solution to the game in pure strategies. In the following example, player 1 has two strategies $S_1 = \{U, D\}$ and player 2 has two strategies $S_2 = \{L, R\}$. Payoffs of the players are given in each situation (U, L), (U, R), (D, L), and (D, R).[1]

[1] The symbols chosen are easy to use: U = Up; D = Down; L = Left; R = Right.

		Player 2	
		L	R
Player 1	U	$u_1(U,L), u_2(U,L)$	$u_1(U,R), u_2(U,R)$
	D	$u_1(D,L), u_2(D,L)$	$u_1(D,R), u_2(D,R)$

A solution of a game is a pair (not necessarily unique) of strategies that a rational pair of players might use. Solutions can be denoted by enclosing a strategy pair between brackets, such as (s_1, s_2) in a two player game for instance, where we will put the strategy adopted by player 1 first. When there is (potentially) more than one decision to be made, the action sets and pure strategy sets are no longer identical. One way of representing a randomizing behavior is to extend the pure strategy concept to a *mixed strategy*. We say that a mixed strategy σ specifies the probability $p(s)$ with which each of the pure strategies $s \in S$ is used. If we suppose the set of pure strategies is $S = \{s_a, s_b, s_c, \ldots\}$, then a mixed strategy can be represented as a vector of probabilities: $\sigma = (p(s_a), p(s_b), p(s_c), \ldots)$.

In this section we enlist techniques for solving games in the normal form and provide examples. A solution is a systematic description of the outcomes that may emerge in a family of games. In strategic games the notion of 'solution' captures a steady state of the play, in which each player holds the correct expectation about the other players' behavior and acts rationally. First of all, it is reasonable to start solving games by eliminating poor strategies for each player by using the so-called *dominance principle*.

Definition 3.3 A strategy for player 1, s_1, **is dominated** by strategy s_1' if for any $s_2 \in S_2$

$$u_i(s_1', s_2) \geq u_i(s_1, s_2)$$

and there exists such $s_2' \in S_2$ that

$$u_i(s_1', s_2') \geq u_i(s_1, s_2')$$

Hence, whatever player 2 does, player 1 is always better off using s_1' rather than s_1. Similarly, we identify domination for player 2.

Some matrix games can be solved by the method of the elimination of dominated strategies. To do so, we have to assume that:

1. The players are rational.
2. The players all know that the other players are rational.
3. The players all know that the other players know that they are rational.
4. …(in principle) ad infinitum.

Consider the following illustrative example.

		Player 2	
		L	R
Player 1	U	(6,2)	(3,3)
	D	(5,1)	(-4,4)

In the given example, for player 1, *U* dominates *D* and, for player 2, *R* dominates *L*. Consequently, we expect that player 1 will not play *D* and player 2 will not play *L*, leaving the solution (*U*, *R*).

Definition 3.4 A **Nash equilibrium** (for two player games) is a pair of strategies $\left(s_1^*, s_2^*\right)$ such that

$$u_1\left(s_1^*, s_2^*\right) \geq u_1\left(s_1, s_2^*\right) \quad \forall s_1 \in S_1$$

and

$$u_2\left(s_1^*, s_2^*\right) \geq u_2\left(s_1^*, s_2\right) \quad \forall s_2 \in S_2.$$

In other words, given the strategy adopted by the other player, neither player could do strictly better (i.e., increase their payoff) by adopting another strategy. Nash equilibrium strategies are the best responses to each other.

Definition 3.5 A solution $\left(s_1^*, s_2^*\right)$ is **Pareto optimal** if no player's payoff can be increased without decreasing the payoff of another player.

Such solutions are also termed (socially) efficient. In general in a game there can be several Pareto optimal outcomes. The Pareto principle tells that to be acceptable as a solution to a game, an outcome should be Pareto optimal. The Pareto principle is a type of *group rationality*. Unfortunately, it does not always hold, as there are games where Pareto optimality comes into direct conflict with an equally cogent principle of *individual rationality*.

Existence of Nash Equilibrium: *Every (two player, two action) game has at least one Nash equilibrium.*

The first existence result was proved by Nash in his famous paper 'Equilibrium Points in n-Person Games', in Proceedings of the National Academy of Sciences, 1950. Nash Theorem (Existence in Finite Games) Finite games, i.e., games with a finite number of players and a finite number of strategies for each player, have a mixed strategy Nash equilibrium.

Consider a (two player, two action) game with arbitrary payoffs:

		Player 2 L	Player 2 R
Player 1	U	(a,b)	(c,d)
Player 1	D	(e,f)	(g,h)

Let us consider pure-strategy Nash equilibria: if $a \geq e$ and $b \geq d$ then (U, L) is Nash equilibrium; if $e \geq a$ and $f \geq h$ then (D, L) is Nash equilibrium; if $c \geq g$ and $d \geq b$ then (U, L) is Nash equilibrium; if $g \geq c$ and $h \geq f$ then (D, L) is Nash equilibrium.

There is no pure strategy Nash equilibrium if either

1. $a < e, f < h, g < c, d < b$
2. $a > e, f > h, g > c, d > b$.

Examples:

- Game with one Nash equilibium (in pure strategies), which is Pareto optimal

		Player 2 L	Player 2 R
Player 1	U	(2,3)	(3,2)
Player 1	D	(1,0)	(0,1)

In this game player 1 always prefers to play *U* no matter what player 2 does. Player 1 strategy *U* dominates strategy *D*, and player 2 knowing that, will choose strategy *L*. Although the payoffs looks symmetric, the arrangement of the payoffs favors player 2. It also appears that the Nash equilibrium (U, L) is Pareto optimal.

- Game with one Nash equilibrium, but the equilibrium is not Pareto optimal

		Player 2 L	Player 2 R
Player 1	U	(3,3)	(-1,5)
Player 1	D	(5,-1)	(0,0)

Here, Nash equilibrium (D, R) is unique and can be found using the principle of dominance: Player 1's *U* is dominated by player 1's *D*, player 2's *L* is dominated by player 2's *R*. However, neither of the players are satisfied with outcome (0, 0), because they both could be better off at (U, L), being (3, 3).

- Game without Nash equilibrium in pure strategies

		Player 2 L	Player 2 R
Player 1	U	(2,4)	(1,0)
Player 1	D	(3,1)	(0,4)

In this example we can see that in every situation in the game, one of the players has an incentive to deviate to another strategy.

- Game with multiple Nash equilibria and, therefore, more than one possible solution

		Player 2 L	Player 2 R
Player 1	U	(1,1)	(2,5)
Player 1	D	(5,2)	(-1,-1)

We can notice that this game has two equilibria (D, L) and (U, R). Moreover, player 1 prefers equilibrium (D, L) and player 2 prefers equilibrium (U, R). Both equilibria are Pareto optimal and if players will decide to get their preferred equilibrium, the situation can end up in the 'undesirable' solution (D, R). Without assumption of additional interaction between the players, game theory cannot give a clear answer as to which equilibria players will go for.

In the following subsections we will consider three types of coordination game, which as we will see further have direct implication into safety decision making in chemical clusters.

Coordination Games: Type 1: Prisoner's Dilemma Game

Let us consider a classic game theory example, called *Prisoner's Dilemma*. The situation involves two criminals, who are questioned by the police in connection with a serious crime. Without a confession, the police only has enough evidence to convict the two criminals on a lesser charge. The police makes the following offer to both prisoners: if one confesses that both committed the serious crime, then the confessor will be set free and the other will spend 5 years in jail (4 for the crime and 1 for obstructing justice); if both confess, then they will each get the 4-year sentence; if neither confess, then they will each spend 2 years in jail for the minor offense. The criminals are held in separate cells so they cannot communicate and do not know about what the other one said.

The game can be described in a tabular form where the possible courses of action open to each prisoner are (1) $C =$ 'not confess' (or 'Cooperate'), (2) $D =$ 'confess' (or 'Deviate'). The payoffs are given in terms of years of freedom lost. The payoffs for the first prisoner (player 1) are given first in each pair of entries in the table; those for the other prisoner (player 2) come second.

		Prisoner 2 (=Player 2) C	Prisoner 2 (=Player 2) D
Prisoner 1 (=Player1)	C	(-2,-2)	(-5,0)
	D	(0,-5)	(-4,-4)

Game-theoretic analysis suggests the following behavior to each prisoner. First, we consider Player 1. Player 1 knows that, if Player 2 decides not to confess, then Player 1 will do better if he confesses the crime because that leads to 0 years in jail rather than 2 years. On the other hand, if Player 2 decides to confess, then Player 1 should also confess because that leads to 4 years in jail rather than 5. So whatever Player 2 does, Player 1 is better off confessing. Similarly, reasoning can be applied to Player 2. So both prisoners should confess.

The interest in this game arises from the following observation. Both players, by following their individual self-interest, end up worse off than if they had kept quiet. One may argue that the criminals should have had an agreement before being arrested that they would not betray. However, when each prisoner has no way of ensuring that the other follows this agreement, the equilibrium outcome would be (Confess, Confess).

		Player 2 C	Player 2 D
Player 1	C	(a,b)	(c,d)
	D	(e,f)	(g,h)

The conditions for the game Nash equilibria and Pareto optimal solution result in the following conditions for players' payoffs in case of a Prisoner's dilemma game:

$$e \geq a \geq g \geq c,$$

$$d \geq b \geq h \geq f.$$

Coordination Games: Type 2: Stag Hunt Game

Suppose that there are only two hunters, and that they must decide simultaneously whether to hunt for stag or for hare. If both hunt for stag, they will catch one stag and share it equally. If both hunt for hare, they both will catch one hare. If one hunts for hare while the other tries to take a stag, the former will catch a hare and the latter will catch nothing. Each hunter prefers half a stag to one hare. Let the hunters be players. Each player has the choice between two strategies: hunt *Stag* and hunt *Hare*. The payoff to their choice is the pray. If, for instance, a stag is worth 4 utils (a 'util' can be considered as a unity for utility) and a hare is worth 1 util, then when both players hunt stag each has a payoff of 2 utils. A player who hunts hare has payoff 1, and a player who hunts stag by himself has payoff 0.

		Hunter 2 (=Player 2) C	D
Hunter 1 (= Player 1)	C	(2,2)	(0,1)
	D	(1,0)	(1,1)

Cooperating—both hunting stag—is Nash equilibrium, in that neither player has a unilateral incentive to change the strategy. Therefore, *Stag* hunting seems like a possible outcome of the game. However, cooperation is not a unique conclusion. If each player believes that the other will hunt hare, each is better off hunting hare himself. Thus the noncooperative outcome—both hunting hare—is also Nash equilibrium. The game has only one Pareto optimal solution (*Stag*, *Stag*). However, game theory cannot give a certain answer as to which of the equilibria will realize. Temptation to play *Hare* can arise if one player doubts the other player's rationality, i.e., willingness to cooperate. Without more information about the context of the game, it is difficult to know which outcome to predict.

Let us consider a general pay-off matrix of a stag hunt game.

		Player 2 C	D
Player 1	C	(a,b)	(c,d)
	D	(e,f)	(g,h)

The conditions for the game Nash equilibria and Pareto optimal solution result in the following conditions for players' payoffs in case of a Stag hunt game:

$$a \geq e \geq g \geq c,$$

$$b \geq d \geq h \geq f.$$

Usually the prisoner's dilemma game is assumed to best represent the problem of social cooperation. Meanwhile, the stag hunt game focuses on an equally interesting perspective on cooperation and its problems (for an overview, see Skyrms 2004).

The stag hunt game can be considered as a game which describes a conflict between safety and social cooperation. When two players trust that the other one will play *Stag*, they will coordinate their strategies and cooperate. In this way, they will reach a Pareto optimal outcome, which is given by full cooperation. If either of the players does not blindly trust another one, the Pareto-inefficient equilibrium (*Hare*, *Hare*) can be a possible outcome of this game.

Coordination Games: Type 3: Chicken Game
Like in the Prisoner's Dilemma game and in the Stag Hunt game, the situation when both players cooperate is Pareto optimal. Similarly to the Stag Hunt game, the Chicken Game is a coordination game with multiple Nash Equilibria. Similarly to Prisoner's Dilemma, in the Chicken Game the situation in which both players cooperate is not stable. Unlike the Prisoner's Dilemma game, the Chicken Game is a coordination game with partial cooperation, i.e., if one of the players does not cooperate then the number of equilibria depends on the number of players, as well as their strategies. Similarly each player, however, prefers a different equilibrium outcome in this game. Moreover, Nash Equilibria in this game differ from the ones in the Stag Hunt game. There are two drivers who are headed toward a single lane bridge from opposite directions. Both drivers have the option to either drive straight ahead across the bridge or to stop and give way to the other driver to (safely) cross the bridge. The driver that stops loses in this game (gets payoff of **–1** util), the driver who drives straight ahead, wins (gets **1** util). If both drivers choose to drive straight ahead, they end up in a disastrous car accident, the worse possible outcome for both drivers (each gets **−5** utils). If both drivers stop and peacefully negotiate, then no one looses his face but also no one wins (each driver gets **0** utils). The pay-offs of this example are given in table below

		Driver 2 (= Player 2) C	Driver 2 (= Player 2) D
Driver 1 (= Player 1)	C	(0,0)	(-1,1)
	D	(1,-1)	(-5,-5)

The game has two Nash equilibria: (Stop, Drive) and (Drive, Stop), both of these equilibria are Pareto optimal, as well as the situation (Stop, Stop). The situation (Stop, Stop) is not stable because either of the players will have incentives to unilaterally deviate. Each player prefers different equilibria to take place. Let us now consider a general pay-off matrix of a Chicken Game to derive general conditions.

		Player 2 C	Player 2 D
Player 1	C	(a,b)	(c,d)
	D	(e,f)	(g,h)

The conditions for the game Nash equilibria and Pareto optimal solutions result in the following conditions for players' payoffs

$$e \geq a \geq c \geq g,$$

$$d \geq b \geq f \geq h.$$

The chicken game presents the third type of coordination games. Here, only partial cooperation is a stable solution, and thus this case represents a middle case between the Stag Hunt game and the Prisoner's Dilemma game.

3.4 Extensive Form Games with a Discrete Set of Strategies

In the previous section we discussed the concept of the 'strategic' (or 'normal') form of a game. As already mentioned, a strategic game is a model of a situation in which each player chooses his or her plan of action once and for all, and all players' decisions are made simultaneously (that is, when choosing a plan of action, each player is not informed of the plan of action chosen by any other player). It is a straightforward method of analysis of games with imperfect information, e.g. when moves of players are simultaneous or unknown until the end. However, the details of many games are lost in such a simple model. The model of an extensive game specifies the possible orders of events; each player can consider his/her plan of action not only at the beginning of the game but also whenever he/she has to make a decision. A general model of an extensive game allows each player, when making his/her choices, to be imperfectly informed about what has happened in the past. However, for the most of this section we focus on games of perfect information, i.e. when the participants are fully informed about each others' moves.

3.4.1 Extensive Games with Perfect Information

An extensive game is a detailed description of the sequential structure of the decision problems encountered by the players in a strategic situation. There is perfect information in such a game if each player, when making any decision, is perfectly informed of all the events that have previously occurred. For simplicity we initially restrict attention to games in which only one player makes decisions at the same time and all relevant moves are made by the players (no randomness ever intervenes). We recall that a game consists of a set of players, the rules of the game, and the payoffs that a player receives as a function of all the moves taken in the game. More formally, the following definition can be given.

Definition 3.6 An **extensive game with perfect information** has the following components:

1. A set of players, N.
2. A directed graph, i.e., a set of nodes X and arrows connecting the nodes. This must form a tree which means:
   - There is a single initial node x^0, i.e., a node with no arrows pointing toward it.
   - For each node, there is a uniquely determined path of arrows connecting it to the initial node (this is also called a *path* to the node).
3. Nodes are divided into
   - Terminal nodes X_T, i.e., with no outward pointing arrows.
   - Decision nodes $X \backslash X_T$, i.e., with outward pointing arrows.
4. Each arrow represents an action available to the player at the decision node at the origin of the arrow (a function $a : X \backslash X_T \to A$). If there is a node from x to x' then we say that x' is a successor of x.
5. A set of sequences H (finite or infinite) that satisfies the following three properties
   - The empty sequence $\emptyset$ is a member of H.
   - If $\left(a^k\right)_{k=1,\ldots,k} \in H$ (where K may be infinite) and $L < K$ then $\left(a^k\right)_{k=1,\ldots,k} \in H$.
   - If an infinite sequence $\left(a^k\right)_k^\infty$ satisfies $\left(a^k\right)_{k=1,\ldots,k} \in H$ for every positive integer L then $\left(a^k\right)_k^\infty \in H$

     Each member of set H is a **history**; each component of a history is an action taken by a player. A history $\left(a^k\right)_{k=1,\ldots,k} \in H$ is terminal if it is infinite or if there is no a^{k+1} such that $\left(a^k\right)_{k=1,\ldots,k} \in H$. The set of terminal histories is denoted Z.
6. Each decision node is labeled as belonging to a player in the game (that is, that player who has to take the decision). A function P that assigns to each non-terminal history (each member of $H \backslash Z$) a member of N. (P is the player function, $P(h)$ being the player who takes an action after a certain history $h \in H$).
7. Each player $i \in N$ has a payoff function $u_i : Z \to R$.

A normal form of a game is a 'deducted' representation because the normal form game can be represented as extensive form where the order of moves are hidden, i.e., in the case where decision nodes for each player belong to one information set.

The structure of an extensive game is specified as a triple $\Gamma = <N, H, P>$. If the set H of possible histories is finite then the game is called 'finite'. If the longest history is finite then the game has a 'finite horizon'. Let h be a history of length k; after any nonterminal history h player $P(h)$ chooses an action from the set $A(h) = \{a : (h, a) \in H\}$. The empty history is the starting point of the game, and

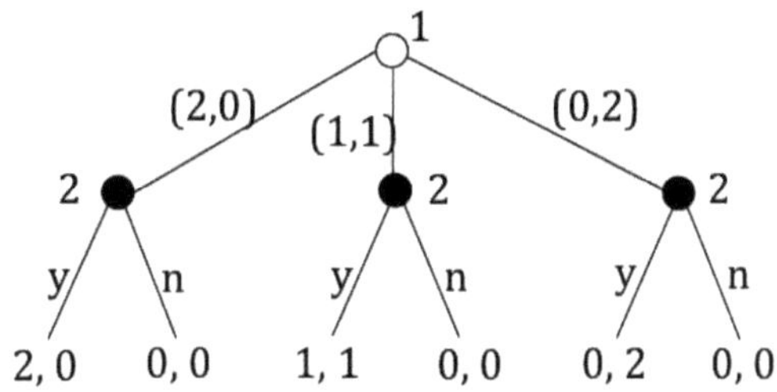

Fig. 3.1 An extensive game that models the procedure for allocating two identical indivisible objects between two individuals

corresponds to the initial node. At this point player $P(\emptyset)$ chooses a member of $A(\emptyset)$. For each possible choice a^0 from this set player $P\ (a^0)$ subsequently chooses a member of the set $A\ (a^0)$; this choice determines the next player to move, and so on. A history after which no more choices have to be made is terminal.

Let us suggest an alternative definition of an extensive game in which the basic component is a tree (a connected graph with no cycles). In this formulation each node corresponds to a history and any pair of nodes that are connected corresponds to an action.

In Fig. 3.1 two players use the following procedure to share two desirable identical indivisible objects. One of them proposes an allocation, which the other, then either accepts or rejects. In the event of rejection, neither person receives either of the objects. Each person cares only about the number of objects he or she obtains.

In the game tree, the node at the top of the diagram is an initial node, which represents the initial history $\emptyset$ (the starting point of the game). It is indicated that $P(\emptyset) = 1$ (player 1 makes the first move). The three arrows that come out from the initial node, correspond to the three members of action set $A(\emptyset)$ (the possible actions of player 1); the labels beside these arrows are the names of the actions, $(k, 2 - k)$,being the proposal to give k of the objects to player 1 and the remaining 2-k to player 2.

Each arrow leads to a small disk beside which is the label 2, indicating decision nodes of player 2, which takes an action after any history of length one. The labels beside the line segments that emanate from these disks are the names of player 2's actions, y (yes) meaning 'accept' and n (no) meaning 'reject'. The numbers below the terminal histories are payoffs that represent the players' preferences. The first number in each pair is the payoff of player 1 and the second is the payoff of player 2.

In this section we have defined the model of an extensive game with perfect information. As in the previous section we need now to specify a winning strategy for players. It appears that the earlier introduced Nash equilibrium is unsatisfactory in this model since it ignores the sequential structure of the decision problems. In the following section we discuss a solution concept, in which a player is required to reassess their plans as the play proceeds, and which is thus more suitable for extensive games with perfect information.

3.4.2 Backward Induction and Subgame Perfect Equilibrium

A strategy of a player in an extensive game is different from a concept of a steady state of strategic games. Here, it is a plan that specifies the action chosen by the player for every history after which it is his/her turn to move. Let us consider an extensive form game with perfect information.

Definition 3.7 A **strategy** of player $i \in N$ in an extensive game with perfect information $\langle N, H, P, (u_i) \rangle$ is a function that assigns an action in $A(h)$ to each nonterminal history $h \in H \backslash Z$ for which $P(h) = i$.

This definition distinguishes between an action and a strategy. A strategy specifies the action chosen by a player for every history after which it is his or her turn to move, even for histories that, if the strategy is followed, are never reached. For each strategy profile $s = (s_i)_{i \in N}$ in the extensive game $\langle <N, H, P(u_i) \rangle$ we define the outcome $O(s)$ of s to be the terminal history that results when each player $i \in N$ follows the precepts of s_i.

To illustrate the notion of a strategy, return to the game in Fig. 3.1. Player 1 takes an action only after the initial history $\emptyset$, so that we can identify each of his/her strategies with one of the three possible actions that he/she can take after this history: (2, 0), (1, 1), and (2, 0). Player 2 takes an action after each of the three histories (2, 0), (1, 1), and (0, 2), and in each case he or she has two possible actions. Thus we can identify each of his/her strategies with a triple $a_2b_2c_2$ where a_2, b_2, and c_2 are the actions that he/she chooses after the histories (2, 0), (1, 1), and (0, 2). The interpretation of player 2's strategy $a_2b_2c_2$ is that it is a contingency plan: if player 1 chooses (2, 0) then player 2 will choose a_2; if player 1 chooses (1, 1) then player 2 will choose b_2; and if player 1 chooses (0, 2) then player 2 will choose c_2.

The extensive game with perfect information $\Gamma = <N, H, P, (u_i)\rangle$ can be presented as the strategic game $<N, (S_i), (u_i)\rangle$ in which for each player $i \in N, S_i$, is the set of strategies of player *1* in Γ, and $ui(s) = ui(O(s))$. The first solution concept we can apply for an extensive game is Nash equilibrium.

Definition 3.8 Nash equilibrium of an extensive game with perfect information $<N, H, P, (u_i)\rangle$ is a strategy profile s^* such that for every player $i \in N$ we have $u_i\left(O\left(s^*_{-1}, s^*_i\right) \geq u_i\left(O\left(s^*_{-1}, s_i\right)\right.\right.$ for every strategy s_i of player i.

Nash equilibrium concept ignores the sequential structure of the game; it treats the strategies as choices that are made once and for all, before the play begins. In this way, Nash equilibrium delivers strategy profiles, which can contain the players' non-credible threat.

Let us introduce a solution concept, in which a player is required to reassess his or her plans as play proceeds. To solve the extensive game, we start from the set of terminal nodes Z_T. For each of the longest nonterminal histories in the game we choose an optimal action for the player whose turn it is to move (notice that the remaining game is a single player decision problem). We replace each of these histories with a terminal history in which the payoff profile is the resulting one when the optimal action is chosen; then we repeat the procedure, working our way back to the start of the game. This process can be summarized as the following algorithm.

Definition 3.9 Process of backward induction

- consider all paths from the initial node x_0 to the last nodes X_T
- the longest path has l arrows
- consider game Γ_l with path l
- in a sequentially rational play, replace the last decision node with an optimal payoff vector $\left(u_1\left(x_T^*\right), \ldots, u_N\left(x_T^*\right)\right)$
- get a new game Γ_{l-1} of length $l-1$
- take Γ_{l-1}, perform the same analysis and get the game Γ_{l-2} of length $l-2$
- continue process iteratively until you get game of length 1, which is again a single player game with a defined solution.

It is necessary to point out here that the process of backward induction assumes the play to be *sequentially rational*. A player's strategy, which is part of a proposed strategy profile for playing the game, is sequentially rational if starting at any decision point for a player in the game (including those that may not be reached if the game is conducted according to the strategy profile), his/her strategy from that point on represents a best response to the strategies of the other players. We say that a play in a game is 'sequentially rational' if at any history h_i, player *1* must choose so as to maximize u_i. Moreover, each player has knowledge that other players are sequentially rational as well.

In the process of backward induction, an action is determined for each decision node. The choice at each decision node is optimal given the play at successor nodes. The process of backward induction suggests that every finite extensive game with perfect information has a subgame perfect equilibrium. We now define the notion of subgame perfect equilibrium, which construction is based on backward induction.

Definition 3.10 The **subgame** of the extensive game with perfect information $\Gamma = \langle N, H, P, (u_i)\rangle$ that follows the history h is the extensive game $\Gamma(h) = \langle N, H|_h, P|_h, (u_i|h)\rangle$, where $H|_h$ is the set of sequences h' of actions for which $(h, h') \in H$, $P|_h$ is defined by $P|_h(h') = P(h, h')$ for each $h' \in H|_h$.

A 'subgame' is a part (sub-tree) of a game tree that satisfies the following conditions: it begins at a decision node (for any player); the player knows all the decisions that have been made until that point in time; the sub-tree contains all the decision nodes that follow the initial node (and no others). In this way, a subgame can be considered as a game of its own.

Definition 3.11 A **subgame perfect equilibrium** of an extensive game with perfect information $\Gamma = \langle N, H, P, (u_i)\rangle$ is a strategy profile s^* such that for every player $i \in N$ and every nonterminal history $h \in H \backslash Z$ for which $P(h) = i$ we have

$u_i\left(O\left(s^*_{-i}|_h, s^*_i|_h\right)\right) \geq u_i(O(s^*_{-i}|_h, s_i|_h)$ for every strategy s_i of player i in the subgame $\Gamma(h)$.

'Subgame perfect Nash equilibrium' is a Nash equilibrium in which the behaviour specified in every subgame is Nash equilibrium for the subgame. The notion of subgame perfect equilibrium eliminates Nash equilibria in which the players' threats are not credible. Note that this applies even to subgames that are not reached during a play of the game using Nash equilibrium strategies.

3.5 Hierarchical Tree-Like Games and Imperfect Information

In Chap. 6 we will consider the optimal allocation problem, in which a coordination center of a chemical cluster (called the Multi-Plant Council, see also Chap. 2) allocates financial resources among the chemical plants to channel their actions as may be desired. At this stage of the book, let us assume that the financial resources to be allocated, equal subsidies. It is clear that the subsidy level conditions are generally different for each plant and the provision of subsidies to one group of plants or another happens according to a certain priority order. Thus, the problem is that of constructing an optimal subsidy program by a coordination center (the MPC) and an optimal binary choice of cooperation by the plants. These kind of problems belong to the class of dynamic games of complete but imperfect information. As in the previous section we keep the assumption that the pay proceeds in a sequence of stages, and the moves in all previous stages can be observed before the next stage begins. However, now we shall allow simultaneous moves within each stage. Such simultaneity of moves within stages means that the game is of imperfect information. Nonetheless, these games share important features with the perfect-information games considered in the previous section. We will analyze the following simple example (Gibbons 1992) of a two-stage game of complete but imperfect information:

- Players 1 and 2 simultaneously choose actions s_1 and s_2 from feasible sets S_1 and S_2, respectively.
- Players 3 and 4 observe the outcome of the first stage, (s_1, s_2), and then simultaneously choose actions s_3 and s_4 from feasible sets S_3 and S_4, respectively.
- Payoff are $u_i(s_1, s_2, s_3, s_4)$ for $i = 1, 2, 3, 4$.

We can solve this game by using an approach in the spirit of backwards induction, but this time the first step in working backwards from the end of the game involves solving a real game (the simultaneous-move game between players 3 and 4 in stage two, given the outcome of stage one) rather than solving a single player optimization problem as in the previous section. To keep things simple, let us assume that for each feasible outcome of the first stage game, (s_1, s_2), the second-stage game that remains between players 3 and 4 has a unique Nash equilibrium, denoted by $(s_3^*(s_1, s_2), s_4^*(s_1, s_2))$. If players 1 and 2 anticipate that the second stage behavior of player 3 and 4 will be given by $(s_3^*(s_1, s_2), s_4^*(s_1, s_2))$, then the first stage interaction between players 1 and 2 amounts to the following simultaneous-move game

- Players 1 and 2 simultaneously choose actions s_1 and s_2 from feasible sets S_1 and S_2, respectively.
- Payoff are $u_i(s_1, s_2, s_3^*(s_1, s_2), s_4^*(s_1, s_2))$ for $i = 1, 2$.

Suppose further that (s_1^*, s_2^*) is the unique Nash equilibrium of this simultaneous-move game. We will call $(s_1^*, s_2^*, s_3^*(s_1^*, s_2^*), s_4^*(s_1^*, s_2^*))$ the subgame-perfect outcome of this two-stage game. This outcome is the natural analog of the backward induction outcome in games of complete and perfect information, and the analogy applies to both positive and unattractive features. Players 1 and 2 should not believe a threat by players 3 and 4 in a sense that they can respond with actions that are not a Nash equilibrium in the remaining second stage game, because when the play actually reaches the second stage, at least one of the players 3 or 4 will not want to carry out such a threat (because it is not a Nash equilibrium of the game that remains at that stage). On the other hand, if player 1 is also player 3, and player 1 does not play s_1^* in the first stage: player 4 may then want to reconsider the assumption that player 3 (i.e. player 1) will play $s_3^*(s_1, s_2)$ in the second stage.

Of course, the described situation can be extended for the case in which the number of players is larger than four, and where there are more levels of hierarchy with a different order of players' moves. In the following chapter we shall consider an application of these game-theoretic models and discuss solution techniques for the problem of enhancing cooperation between chemical plants, and specify particular features and conditions of the coordinated actions.

3.6 Games with Incomplete Information

So far we have been assuming that everything in the game was common knowledge for everybody playing. But in fact players may have private information about their own payoffs, about their type or preferences, etc. The way to modeling this situation of asymmetric or incomplete information is by recurring to an idea generated by Harsanyi (1967). The key is to introduce a move by the Nature, which transforms the uncertainty by converting an incomplete information problem into an imperfect information problem. The idea is that the Nature moves, determining players' types: Type is a concept that embodies all the relevant private information about them (such as payoffs, preferences, beliefs about other players, etc.).

In a Bayesian game, the incompleteness of information means that at least one player is unsure of the type (and so the payoff function) of another player.

The normal form representation of a non-Bayesian game with perfect information is a specification of the strategy spaces and payoff functions of players. A strategy for a player is a complete plan of action that covers every contingency of the game, even if that contingency can never arise. The strategy space of a player is thus the set of all strategies available to a player. A payoff function is a function from the set of strategy profiles to the set of payoffs (normally the set of real numbers), where a strategy profile is a vector specifying a strategy for every player.

In a Bayesian game, it is necessary to specify the strategy spaces, the type spaces, the payoff functions and the beliefs for every player. A strategy for a player is a complete plan of action that covers every contingency that might arise for every type that the player might be. A strategy must not only specify the actions of the player given the type that he is, but must specify the actions that he would take if he were of another type. Strategy spaces are defined as above. A type space for a player is just the set of all types.

Definition 3.12 A *Bayesian Game* is a game in normal form with incomplete information that consists of:

1. Players $i \in \{1, 2, \ldots, N\}$
2. Finite action set for each player $a_i \in A_i$
3. Finite type set for each player $\theta_i \in \Theta_i$
4. A probability distribution over types $p(\theta)$(common prior beliefs about the players' types)
5. Utilities $u_i : A_1 \times \ldots \times A_N \times \Theta_1 \times \ldots \times \Theta_N \rightarrow R$.

It is important to briefly discuss each part of the definition.

Types of a player contain all relevant information about certain private characteristics of the player. The type θ_i is only observed by player i, who uses this information both to make decisions and to update his/her beliefs about the likelihood of opponents' types (using the conditional probability $p(\theta_{-i}|\theta_i)$.

Combining actions and types for each player, it is possible to construct the strategies. Strategies will be given by a mapping from the type space to the action space, $s_i : \Theta_i \rightarrow A_i$, with elements $s_i(\theta_i)$. In other words, a strategy may assign different actions to different types.

Finally, each player calculates utilities u_i by taking expectations over types using his/her own conditional beliefs about opponents' types. Hence, if player i uses the pure strategy s_i, other players use the strategies s_{-i} and the player i's type is θ_i, the expected utility can be written as

$$Eu_i(s_i|s_{-i}, \theta_i) = \sum_{\theta_{-i} \in \Theta_{-i}} u_i(s_i, s_{-i}(\theta_{-i}), \theta_i, \theta_{-i})p(\theta_{-i}|\theta_i).$$

A *Bayesian Nash Equilibrium* (BNE) is basically the same concept as a Nash Equilibrium with the addition that players need to take expectations over opponents' types. Hence:

Definition 3.13. *A BNE is a strategy profile* $\mathbf{s}^*$ *such that for every player* i *and every type* θ_i,

$$Eu_i(s_i^*|s_{-i}^*, \theta_i) \geq Eu_i(s_i|s_{-i}^*, \theta_i),$$

for all $s_i(\theta_i) \in S_i$ and for all types θ_i occurring with positive probability.

Thus, a BNE is a Nash Equilibrium of a Bayesian Game. The following proposition holds.

Proposition *Every finite Bayesian Game has a BNE.*

Here are some examples. In all of them, one should take care to go through the necessary formalities of the associated Bayesian game.

Example 1 An opponent of unknown strength.

Consider a two-person game, between the row player (Row) and the column player (Column). The row player's type is known, but the column player can be "strong" (with probability α) or "weak" (with probability $1 - \alpha$). If the column player is strong, the situation is

		Column	
		Fight	Yield
Row	Fight	*-1,1*	*1,0*
	Yield	*0,1*	*0,0*

When the column player is weak, the situation is

		Column	
		Fight	Yield
Row	Fight	*1,-1*	*1,0*
	Yield	*0,1*	*0,0*

To solve this game, let's first note that if the Column is strong, it is a dominant strategy for the Column to play Fight (remark that payoffs in the figure first mention those of Row, second those of Column). If Row fights, he/she gets 1 if the opponent is weak, and, by the dominance argument just made, he/she gets -1 if the opponent is strong. So Row's expected return from fighting is $-\alpha + (1 - \alpha)$. Row's expected return from playing Yield is always 0.

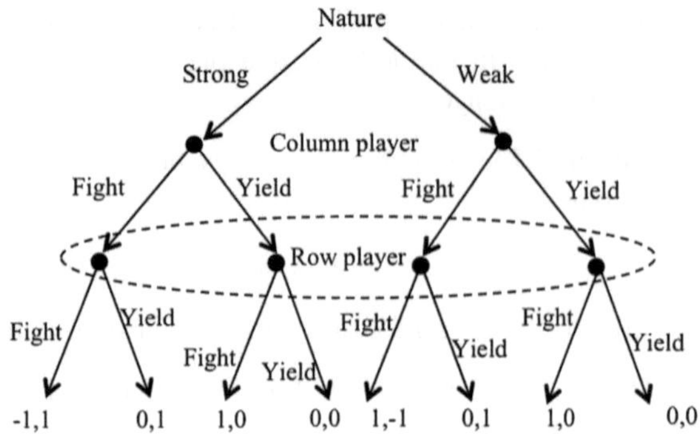

Fig. 3.2 Tree representation of Bayesian game

It follows that if $\alpha > 1/2$, Row will Yield so that both types of Column Fight. On the other hand, if $\alpha < 1/2$, Row will Fight anyway so that the weak type of Column will not fight (Fig. 3.2).

Example 2 Adverse selection.

A firm A is taking over another firm B. The true value of B is known to B but unknown to A. It is only known by A to be uniformly distributed on interval [0, 1]. It is also known that B's value will flourish under A's ownership: it will rise to λx, where x is the pre-takeover value and $\lambda > 1$. All of this description is, in addition, common knowledge.

A strategy for A is a bid y. A strategy for B is a yes–no decision as a function of its value x. So if the type (B, x) accepts, A gets $\lambda x - y$, while (B, x) gets y. If (B, x) rejects, A gets 0 while (B, x) gets x. It follows, therefore, that (B, x) will accept if $x < y$, while he/she will reject if $x > y$. Therefore, the expected value of B in A's eyes conditional on B accepting the offer, is $y/2$. It follows that the overall expected payoff to A is

$$\lambda(y/2) - y = y\left[\frac{\lambda}{2} - 1\right].$$

Therefore the acquirer cannot buy unless the takeover target more than doubles in value after the takeover.

References

Gibbons R (1992) A primer in game theory. Pearson Education Limited, Harlow, England

Harsanyi JC (1967) Games with incomplete information played by "Bayesian" players, part I–III. Part I: the basic model. Manage Sci 14:159–182

Nash J (1950) Equilibrium points in n-person games. Proc Natl Acad Sci 36(1):48–49

Skyrms B (2004) The stag hunt and the evolution of social structure. Cambridge University Press, Cambridge

Von Neumann J, Morgenstern O (1944) Theory of games and economic behavior. Princeton University Press, Princeton

Chapter 4
A Game-Theoretic Model for Cross-Plant Prevention in a Chemical Industrial Park

4.1 Introduction: The Domino Effect Game Model

In this section we introduce the game-theoretic model for cross-plant prevention and describe players' objective parameters. Consider n chemical plants (players) composing a chemical park characterized by

- the probability $0 \leq P_{\text{ii}} \leq 1$ that company i's lack of action can lead to an internally induced loss L_i, $L_i \geq 0$;
- the probability $0 \leq P_{\text{ji}} \leq 1$ that company j's lack of action can cause an externally induced loss L_i to the company i, $j \neq i = 1, \ldots, n$;
- the investment $c_i \geq 0$ into cross-plant prevention, which not only secures company i from occurrence of a major accident with escalation potential (i.e., an internal domino effect), but also guarantees no cross-border effect from i to any other company in the cluster;
- a discrete choice of strategy A_i that can take values Invest (I) or Not Invest (NI) into cross-plant prevention of company i.

A loss caused by an internal domino effect is called 'a direct loss', whereas a loss to other companies (caused by an external domino effect) is considered as 'an indirect impact' and is referred to as 'an indirect loss'. Let $l_i(\{\psi\}, A_i)$ be the expected indirect loss to a company i when it chooses a strategy A_i and ψ is a set of companies in the chemical cluster which choose strategy I ($\psi \subseteq \varphi$). Consider a one-stage game $\Gamma = \{A_1, \ldots, A_n; u_1, \ldots, u_n\}$, where u_i $(i = 1, \ldots, n)$ is cost (negative payoff) for plant (player) i. The cost for company (player) i when it chooses strategy $A_i = \text{I}$, given companies(/players) from a coalition ψ do so as well, and the rest of the companies (players) $\varphi \backslash \psi$ choose NI, is:

$$u_i^{\psi} = c_i + l_i(\{\psi\}, \text{I}) \tag{4.1}$$

If company (player) i's strategy is $A_i = \text{NI}$, then company (player) i's expected cost is:

G. Reniers and Y. Pavlova, *Using Game Theory to Improve Safety within Chemical Industrial Parks*, Springer Series in Reliability Engineering,
DOI: 10.1007/978-1-4471-5052-7_4,

$$u_i^{\varphi\backslash\psi} = L_i P_{\mathrm{ii}} \prod_{j \neq i, j \in \varphi\backslash\psi} (1 - P_{\mathrm{ji}}) + l_i(\{\psi\}, \mathrm{NI})(1 - P_{\mathrm{ii}}) \tag{4.2}$$

In expression (4.2) the first term is an expected direct loss of the company i, and the second term is its expected indirect loss. To make a prediction about the strategy each player chooses, we determine the Nash equilibrium of the game and find out whether cooperation among companies belonging to a chemical cluster is a stable outcome of the game.

Remark Throughout the book, we assume that the situation in which all players invest in domino effect prevention, is a social optimum (or a Pareto optimal solution), and that the situation in which none of the players invests in preventive measures, cannot be socially optimal.

Plants' decisions to invest in domino effects prevention to decrease internal domino risks, also decrease external (cross-plant) domino risks experienced by other plants within the cluster. Hence, the more plants that invest in domino prevention, the higher are the positive externalities in the system. It should be noted that if a player i has decided to invest in domino effect prevention and a player j has to decide whether or not to do likewise, then the higher the probability that j can benefit from preventive investment by i, the less likely it is that j will follow suit and invest as well. The game-theoretic model for cross-plant prevention in a chemical cluster considered in this book is a game with multiple equilibria. As we show further, realized Nash equilibria will depend on the relationship between the model parameters in different types of coordination games.

4.2 Coordination Games of Two Players in Application to the Chemical Industry

We have already introduced players' payoffs in a general form. In this section, we are going to analyze the applicability of the different types of coordination games considered in Chap. 3, where game-specific equilibrium conditions were given. Let us consider the game $\Gamma = \{A_1, A_2; u_1, u_2\}$ between two players. Each player i's payoff is negative and is given as cost function, $u_i(A_1, A_2) = -c_i(A_1, A_2)$ $(i = 1, 2)$. Here, the cost $c_i(A_1, A_2)$ is positive for $i = 1, 2$ and the values are given in Table 4.1. In such a case, the players will solve the dual expenditure minimization problem and the Nash equilibrium profile (A_1^*, A_2^*) is such that for every i

$$c_1(A_1^*, A_2^*) \leq c_1(A_1, A_2^*)$$

and

$$c_2(A_1^*, A_2^*) \leq c_2(A_1^*, A_2),$$

Table 4.1 General cost matrix of player 1 and player 2

Player 1\ player 2	Invest	Not invest
Invest	c_1; c_2	$c_1 + P_{21}L_1$; $P_{22}L_2$
Not invest	$P_{11}L_1$; $c_2 + P_{12}L_2$	$P_{11}L_1(1 - P_{21}) + P_{21}L_1(1 - P_{11})$; $P_{22}L_2(1 - P_{12}) + P_{12}L_2(1 - P_{22})$

where

$$\min_{A_i} c_i\left(A_1^*, A_2^*\right).$$

We assume that investments in safety measures to prevent external domino accidents are completely effective. This means that the probability of this type of accident initiated at the investing player reduces to zero. It should be noted that in reality, the 'zero accident' concept, or 'absolute safety', does not exist: however small the event's probability, it still exists. When a certain player (e.g., player 1) decides to invest in prevention measures, it evaluates the costs of preventive measures as 'c'. Player 2's options still remain 'to invest' or 'to not invest'. On the one hand, when player 2 decides to invest as well, its investment cost is $c_{2.}$ Player 1's investment cost, in this case, it is $c_{1.}$ When player 2, on the other hand, decides not to invest while player 1 invests, it risks a loss of $P_{22}L_2$. In this case, player 1 faces the risk that an external domino accident initiated at player 2 causes damage to its infrastructure. The loss associated with this risk is $P_{21}L_1$. In other words: $P_{21}L_1$ represents the loss for player 1 (L_1) multiplied by the probability that an accident initiated in player 2 will incur damage to player 1, i.e., P_{21}. From now on, we refer to $P_{21}L_1$ as the expected loss for player 1 of an accident initiated inside player 2. For player 2 we find a similar expected loss (i.e., $P_{12}L_2$). The total investment cost for player 1 in case player 2 decides not to invest is therefore, the sum of their direct investment cost and the expected loss of an accident initiated at player 2 (i.e. $c_1 + P_{21}L_1$). In the exact same way, we can present the cost functions for both players when player 2 invests and player 1 decides not to invest.

In a situation where none of the involved companies invest, there is no direct investment cost. Both players only face implicit costs, i.e., costs that only occur when an actual accident occurs. We assume that a loss amounts to the total intrinsic value (i.e., the theoretic or book value) of the plant because an external domino accident is completely effective. This means that in case an accident takes place the plant (not the entire organization but the local company belonging to the organization) will go bankrupt. In other words, an external domino accident destroys the entire plant and can therefore only occur once. Analogous to Reniers et al. (2009), the cost for player 1 in case neither of the players invest is $P_{11}L_1(1 - P_{21}) + P_{21}L_1(1 - P_{11})$. This cost is composed of the sum of the expected loss from an external domino accident initiated within the plant itself conditioned on there being no accident from plant 2 onto plant 1 and the expected loss from an external domino accident initiated at plant 2 onto plant 1 conditioned on there being no accident initiated within the plant itself. These conditions result from the fact that a chemical installation can only explode once and from the assumption that internal and external accidents do not occur at the same time. Players' costs are presented in Table 4.1. In Table 4.1, strategies of player 1 are given in the first column, and strategies of player 2 are given in the first row. First, we write player 1's cost, and then we write player 2's cost.

In the Prisoner's Dilemma game the non-cooperative strategy Not Invest is dominant. This results in the Pareto inferior solution (NI, NI). In a two-player Stag Hunt game, there are two Nash Equilibria. The first equilibrium results if both players fully cooperate (Invest in cross-company prevention measures). The second equilibrium results if both players do not cooperate (Not Invest in prevention measures). However, contrary to the Prisoner's Dilemma game there is no dominant strategy in this game. The former equilibrium is also a Pareto optimal solution. It implies that it is financially best for both players to cooperate and to invest in prevention measures rather than to not cooperate and not invest in prevention measures. In theory, players should thus follow a cooperative strategy to achieve an outcome associated with a stable and Pareto efficient solution. In reality, the cooperative strategy is not dominant and trust-lacking players may well end up in a non-cooperative equilibrium. In Chap. 5, it is demonstrated how the Multi-Plant Council can help to resolve this issue by providing incentives (either granted or imposed) in order to establish the socio-economic optimum, and Chap. 6 will further elaborate on this MPC inducement of cooperation. In a Chicken game, there are two Nash equilibria, but in contrast to the Stag Hunt game, each player prefers a different equilibrium in this game. Equilibrium conditions for the Prisoner's Dilemma game and the Stag Hunt game, and applied to a chemical cluster composed of two chemical plants, are presented in Table 4.2.

In the Prisoner's Dilemma game (see Table 4.2) the condition $P_{\mathrm{ii}}L_i(1-P_{\mathrm{ji}})+P_{\mathrm{ji}}L_i(1-P_{\mathrm{ii}})\leq c_i+P_{\mathrm{ji}}L_i,\ i\neq j=1,2$, can be rewritten as $c_i\geq P_{\mathrm{ii}}L_i(1-2P_{\mathrm{ji}})$, which is a necessary condition for the strategy profile (NI, NI) to be Nash equilibrium. The same condition holds for the Stag Hunt game.

The main difference between the Prisoner's Dilemma game and the Stag Hunt game is realized through the inequality

$$c_i^{\mathrm{SH}}\leq P_{\mathrm{ii}}L_i\leq c_i^{\mathrm{PD}}, \tag{4.3}$$

where the upper index SH denotes Stag Hunt case, and PD denotes Prisoner's Dilemma case. This condition shows that if the investment costs are equal or larger than the expected direct loss from not investing in preventive measures, then the game is of the Prisoner's Dilemma type (Not Invest becomes a dominant strategy)

Table 4.2 Equilibrium conditions for Prisoner's Dilemma game and Stag Hunt game

Type of game	Equilibrium condition
Prisoner's Dilemma	$P_{\mathrm{ii}}L_i\leq c_i\leq P_{\mathrm{ii}}L_i(1-2P_{ji})+P_{\mathrm{ji}}L_i\leq c_i+P_{\mathrm{ji}}L_i$
Stag Hunt	$c_i\leq P_{\mathrm{ii}}L_i(1-2P_{\mathrm{ji}})+P_{\mathrm{ji}}L_i\leq c_i+P_{\mathrm{ji}}L_i\leq P_{\mathrm{ii}}L_i$

is the only equilibrium (no cooperation). If investment costs are smaller or equal to expected direct loss from not investing in preventive measures, then the strategy profile (I, I) is a Nash equilibrium and cooperation between the companies is possible. However, even though (I, I) is a Nash equilibrium, the strategy Invest of either of the players is not necessarily dominant (Stag Hunt game). These observations are summarized in the following proposition:

Proposition 4.1 *Consider a* 2×2 *game* $\Gamma = \{A_1, A_2; u_1, u_2\}$ *of cross-plant prevention in a chemical cluster between two identical players with expected cost functions given as* (4.1) *and* (4.2).

Sufficient conditions of a Prisoner's Dilemma game:

$$c_i^{\mathrm{PD}} \geq P_{\mathrm{ii}} L_i, \quad i = 1, 2. \tag{4.4}$$

Sufficient conditions of a Stag Hunt game:

$$\begin{aligned} c_i^{\mathrm{SH}} &\leq P_{\mathrm{ii}} L_i, & i &= 1, 2, \\ c_i^{\mathrm{SH}} &\geq P_{\mathrm{ii}} L_i (1 - 2P_{\mathrm{ji}}), & i &\neq j = 1, 2. \end{aligned} \tag{4.5}$$

Proof Conditions (4.4) and (4.5) are obtained from Table 4.2. Conditions (4.4) and (4.5) are a helpful tool to identify whether a game belongs to the Stag Hunt type of games or to the Prisoner's Dilemma type of games. The second condition in (4.5) is required to guarantee that the game has no dominant strategies and thus both cooperative and non-cooperative equilibria are possible. If it holds that

$$c_i \leq P_{\mathrm{ii}} L_i (1 - 2P_{\mathrm{ji}}), \quad i \neq j = 1, 2,$$

then there is no conflict in the game and the unique equilibrium is when both players invest. □

In most cases regarding the game-theoretic model for cross-plant prevention, we are interested in finding equilibria in pure strategies. The following proposition suggests that in a game between two players with identical objective functions, there always exists at least one equilibrium in pure strategies. This proposition also holds when the number of players is larger than two.

Proposition 4.2 *Consider a* 2×2 *game* $\Gamma = \{A_1, A_2; u_1, u_2\}$ *of cross-plant prevention between two identical players with expected cost functions given as (1) and (2). Such a game (if solved in pure strategies) can have only one of the following equilibria:*

1. *One Nash equilibrium, which is not socially optimal (PD)*
2. *One Nash equilibrium, which is socially optimal ('no conflict' situation)*
3. *Two equilibria, one of which is socially optimal (SH).*

Proof In Proposition 4.1 we have shown how equilibria conditions of the Stag Hunt game and the Prisoner's Dilemma game are realized for the considered game of chemical plants. A 'no conflict' situation means that both players have 'Invest' as their dominant strategy and the game has a Nash equilibrium when both players play Invest.

For the coordination game of the type 'Chicken Game', equilibria conditions are given as follows:

$$P_{\text{ii}}L_i \leq c_i \leq c_i + P_{\text{ji}}L_i \leq P_{\text{ii}}L_i(1 - 2P_{\text{ji}}) + P_{\text{ji}}L_i, \quad i \neq j = 1, 2. \tag{4.6}$$

Notice that from conditions (4.6) it follows that

$$c_i \leq L_i P_{\text{ii}}(1 - 2P_{\text{ji}}) \leq P_{\text{ii}}L_i, \quad i \neq j = 1, 2, \tag{4.7}$$

which contradicts the first condition of the Chicken Game (see Chap. 3, Sect. 3.3, for the general conditions of Chicken Game) that

$$P_{11}L_1 \leq c_1.$$

This observation obviously rules out the Chicken Game option as a coordination game case for our game with chemical plants. □

If the players' objective functions are not identical, the range of the game outcomes is richer. Proposition 4.3 summarizes the results of Proposition 4.2 for all possible outcomes of the considered coordination game.

Proposition 4.3 *Consider a* 2×2 *game* $\Gamma = \{A_1, A_2; u_1, u_2\}$ *of two non-identical players with expected cost functions given as* (4.1) *and* (4.2). *The game can have one of the following solutions in pure strategies:*

1. *Unique Nash equilibrium, which is not socially optimal, i.e., (NI, NI), (I, NI), or (NI ,I),*
2. *Unique Nash equilibrium, which is socially optimal, i.e., (I, I),*
3. *Two equilibria, one of which is socially optimal, i.e., (I, I) and (NI, NI).*

Proof We have shown in Proposition 4.2 that the given model has the following properties: each player can either play Invest as a dominant strategy, or play Not Invest a dominant strategy; or play Invest if another player invests, and play Not Invest, if another player does not invest. This observation leads to the following possible situations.

Let player 1 have Not Invest as a dominant strategy. If player 2 has a dominant strategy Invest, then the game has a unique Nash equilibrium (NI, I), which is not socially optimal. If player 2 has a dominant strategy Not Invest, then the game has a unique Nash equilibrium (NI, NI), which is not socially optimal either (Prisoner's Dilemma game). If player 2 plays Invest if another player plays Invest, and player 2 plays Not Invest if another player plays Not Invest (as in a Stag Hunt game), then the unique Nash equilibrium is (NI, NI).

Let player 1 have Invest as a dominant strategy. If player 2 has a dominant strategy Invest, then the game has a unique Nash equilibrium (I, I), which is socially optimal and corresponds to a conflict situation. If player 2 has a dominant strategy Not Invest, then the game has a unique Nash equilibrium (I, NI), which is not socially optimal. If player 2 plays Invest if another player plays Invest, and player 2 plays Not Invest if another player plays Not Invest (as in a Stag Hunt game), then the unique Nash equilibrium is (I, I).

Let player 1 play Invest if another player plays Invest, and player 2 plays Not Invest if another player play Not Invest (as in Stag Hunt game). If player 2 has a dominant strategy Invest, then the game has a unique Nash equilibrium (I, I), which is socially optimal. If player 2 has a dominant strategy Not Invest, then the game has a unique Nash equilibrium (NI, NI). If player 2 plays Invest if another player plays Invest, and player 2 plays Not Invest if another player plays Not Invest (as in a Stag Hunt game), then the game has two Nash equilibria in pure strategies, that is (I, I) and (NI, NI). □

It is useful to notice that cases when only one of both players invests and the other one does not, become possible in the game where players' objective functions are not identical. Although these outcomes are not socially optimal, they nevertheless indicate that some investment will take place. As well as Proposition 4.2, results of Proposition 4.3 can be extended to a number of players larger than two.

4.3 Welfare Effect

In this section, we consider the players' gains from cooperating in the Prisoner's Dilemma game and in the Stag Hunt game.

Condition (4.3) suggests that condition $P_{ii}L_i \leq c_i$ holds for Prisoner's Dilemma games. Given L_i and P_{ii}, given also the payoff structure in Table 4.1 and condition (4.3), plants' welfare in the situation (I, I) and (NI, NI) is given as follows (remember that u_i $(i = 1, \ldots, n)$ is cost (negative payoff) for player i):

$$0 < u_i(\mathrm{NI}, \mathrm{NI}) - u_i(\mathrm{I}, \mathrm{I}) \leq L_i P_{ji}(1 - 2P_{ii}). \tag{4.8}$$

The welfare benefit from cooperation for plant i cannot be larger than expected indirect losses $L_i P_{ji}$. The larger the probability of an internal accident, the smaller the benefits of cooperation.

From Table 4.2 and condition (4.3) it follows that in the Stag Hunt game $P_{ii}L_i(1 - 2P_{ji}) \leq c_i \leq P_{ii}L_i$. Consequently, the interval for the values of the cost parameter c_i has length $2L_i P_{ii} P_{ji}$, and can become very narrow for a small probability of a direct effect, that is P_{ii}, as well as of an indirect effect, that is P_{ii}. Given L_i and P_{ii}, the length of the interval for c_i depends only on P_{ji}, which represents the probability of an indirect effect of plant j on plant i. Based on the estimations of c_i in the Stag Hunt game and the payoff expressions in Table 4.1, we can compare the plants' welfare in the situations (I, I) and (NI, NI):

$$L_i P_{ji}(1 - 2P_{ii}) \leq u_i(\mathrm{NI}, \mathrm{NI}) - u_i(\mathrm{I}, \mathrm{I}) \leq L_i P_{ji}. \tag{4.9}$$

As we can notice, the welfare benefits from cooperation for plant i are at least as large as $L_i P_{\rm ji}(1-2P_{\rm ii})$ but not larger than the expected indirect losses $L_i P_{\rm ji}$, and thus limited. The larger the probability of an internal accident is, the smaller the benefit from cooperation becomes.

Proposition 4.4 *In a game-theoretic model for cross-plant prevention in a chemical cluster, investment cost in preventive measures in a Prisoner's Dilemma game is at least as large as in a Stag Hunt game, and gains from cooperation in a Stag Hunt game are at least as large as in a Prisoner's Dilemma game.*

Proof The proof of the proposition follows from the analysis presented above. According to the conditions in (4.3) for the Stag Hunt game and the Prisoner's Dilemma game, it holds that

$$c_i^{\rm SH} \le L_i P_{\rm ii} \le c_i^{\rm PD}.$$

Combining (4.8) and (4.9), gains from cooperation for the considered games are

$$0 < u_i^{\rm PD}({\rm NI},{\rm NI}) - u_i^{\rm PD}({\rm I},{\rm I}) \le L_i P_{\rm ji}(1-2P_{\rm ii}) \le u_i^{\rm SH}({\rm NI},{\rm NI}) - u_i^{\rm SH}({\rm I},{\rm I}) \le L_i P_{\rm ji}.$$

We can expect that players involved in a Stag Hunt game can jointly benefit from cooperating, more than players involved in a Prisoner's Dilemma game; however, the total gains difference is not more than the expected indirect damage $L_i P_{\rm ji} + L_j P_{\rm ij}$, $i \ne j = 1, 2$. □

4.4 Bayesian Coordination Games

Consider two chemical plants composing a multi-plant area. Plant 1 and Plant 2 are considering the possibility of avoiding domino-type accidents by investing in preventive measures. The decision to invest or not depends (at least partially) on the other player's decision. Assume that the investment costs of Plant 1 are either low or high and that Plant 1 knows its costs but Plant 2 does not. Plant 2 has beliefs such that Plant 1's cost is $\overline{c}_1$ with high probability ρ and $\underline{c}_1$ with low probability $1-\rho$, see Table 4.3. Otherwise the total expected cost structure is the same as in Table 4.1.

Plant 2 has no private information. Plant 1 has private information about its costs and its type space has two elements, $\overline{c}_1$ and $\underline{c}_1$. A strategy for Plant 1 is an action for each of the two types, $s_1(c) \in A_1 = \{{\rm I}, {\rm NI}\}$, where $c \in \Theta_1 = \{\overline{c}_1, \underline{c}_1\}$. A strategy for Plant 2 is a single action $s_2 = a_2 \in A_2 = \{{\rm I}, {\rm NI}\}$.

As we have shown in the previous sections, player 1's strategic behavior depends on the cost structure and can be summarized as follows depending on the model parameters

$$c_1^I \le L_1 P_{11}(1-2P_{21}) \le c_1^{\rm SH} \le L_1 P_{11} \le c_1^{\rm PD},$$

where index I corresponds to 'no conflict' behavior where player i always prefers Invest, index SH corresponds to the Stag Hunt game where player i plays Invest if

Table 4.3 Bayesian investment game between player 1 and player 2

Player 1\ player 2	Invest	Not invest
High investment cost $\bar{c}_1$ $[\rho]$		
Invest	$\bar{c}_1$; c_2	$\bar{c}_1 + P_{21}L_1$; $P_{22}L_2$
Not invest	$P_{11}L_1$; $c_2 + P_{12}L_2$	$P_{11}L_1(1-P_{21}) + P_{21}L_1(1-P_{11})$; $P_{22}L_2(1-P_{12}) + P_{12}L_2(1-P_{22})$
Low investment cost $\underline{c}_1$ $[1-\rho]$		
Invest	$\underline{c}_1$; c_2	$\underline{c}_1 + P_{21}L_1$; $P_{22}L_2$
Not invest	$P_{11}L_1$; $c_2 + P_{12}L_2$	$P_{11}L_1(1-P_{21}) + P_{21}L_1(1-P_{11})$; $P_{22}L_2(1-P_{12}) + P_{12}L_2(1-P_{22})$

player j invests and vica versa, index PD corresponds to the Prisoner's Dilemma game where Not Invest is a dominant strategy.

We are going to consider two coordination games with incomplete information. In both games, we assume that player 2's expected cost function profile is that of the Stag Hunt game: player 2 does not have a dominant strategy, it prefers Invest if player 1 invests and Not Invest if player 1 plays Not Invest.

Bayesian game 1. The game is presented in Table 4.3. We assume that if player 1 has high costs, Not Invest becomes its dominant strategy as in the Prisoner's Dilemma game, but in case it has a low cost $\underline{c}_1$ then Invest is its dominant strategy as in a no-conflict situation. The Bayesian Nash equilibrium can be written in the form of a triple:

$$((s_1(\bar{c}_1), s_1(\underline{c}_1)), s_2).$$

We list the strategy of player 1 between brackets for the high- and low-cost types, respectively, and the third element in the triple is player 2's strategy. Since the high-cost player 1 has a dominant strategy Not Invest and the low-cost player 1 has a dominant strategy Invest, any Bayesian equilibrium s should be such that $s_1(\bar{c}_1) = \text{NI}$ and $s_1(\underline{c}_1) = \text{I}$. Player 2 takes this into account, as well as the ex ante probability distribution over types of player 1 (that is, probability ρ that player 1 has a high cost type, and probability $1-\rho$ of player 1 having a low cost type). Player 2's expected total cost if it chooses to invest, is:

$$u_2(\text{I}) = c_2 + \rho P_{12}L_2,$$

Player 2's payoff if it chooses Not Invest is

$$u_2(\text{NI}) = P_{22}L_2 + \rho L_2 P_{12}(1 - 2P_{22}).$$

Moreover, player 2 weakly prefers Invest if

$$\rho \leq \tilde{\rho} := \frac{L_2 P_{22} - c_2}{2L_2 P_{12} P_{22}}.$$

Table 4.4 Example of Bayesian investment cost game between player 1 and player 2

Player 1\ player 2	Invest	Not invest
High investment cost $\bar{c}_1$ $[\rho]$		
Invest	(15, 19.8)	(20, 20)
Not invest	(10, 39.8)	(14.8, 39.2)
Low investment cost $\underline{c}_1$ $[1-\rho]$		
Invest	(8, 19.8)	(13, 20)
Not invest	(10, 39.8)	(14.8, 39.2)

Invest is a best response to the decision rule followed by player 1 as long as $\rho \leq \tilde{\rho}$ and Not Invest is a best response if $\rho > \tilde{\rho}$. Thus, if $\rho \leq \tilde{\rho}$, the Bayesian-Nash equilibrium is given by $((\text{NI}, \text{I}), \text{I})$, and if $\rho > \tilde{\rho}$, the Bayesian-Nash equilibrium is given by $((\text{NI}, \text{I}), \text{NI})$.

To illustrate how Bayesian-Nash equilibria can be found for the 'Bayesian game 1', we assume the following illustrative model parameters: $\bar{c}_1 = 15, \underline{c}_1 = 8, c_2 = 19.8, P_{21} = 0.01, P_{12} = 0.02, P_{11} = 0.02, P_{22} = 0.02; L_1 = 500, L_2 = 1000.$ Table 4.4 further elaborates and illustrates the hypothetical example.

Player 2's expected total cost if it chooses to invest is:

$$u_2(\text{I}) = 19.8 + 20\rho,$$

Player 2's payoff if it chooses Not Invest is:

$$u_2(\text{NI}) = 20 + 19.2\rho.$$

Player 2 weakly prefers Invest if:

$$\rho \leq \tilde{\rho} := \frac{1}{4}.$$

So Invest is a best response to the decision rule followed by player 1 as long as $\rho \leq \frac{1}{4}$ and Not Invest is a best response if $\rho > \frac{1}{4}$. Thus if $\rho \leq \frac{1}{4}$, the Bayesian-Nash equilibrium is given by $((\text{NI}, \text{I}), \text{I})$ and if $\rho > \frac{1}{4}$ the Bayesian-Nash equilibrium is given by $((\text{NI}, \text{I}), \text{NI})$.

Bayesian game 2. Assume as before that player 2's expected cost function profile is that of the Stag Hunt game (that is, player 2 does not have a dominant strategy, he prefers Invest if player 1 invests and Not Invest if player 1 plays Not Invest). In the second Bayesian game we assume that if player 1 has a low cost $\underline{c}_1$ then their payoff structure is the same as in the Stag Hunt game, but in case he has high costs $\bar{c}_1$, Not Invest becomes their dominant strategy, like in a Prisoner's Dilemma game (see Table 4.5).

Since the high-cost player 1 has a dominant strategy Not Invest, any Bayesian equilibrium s should be such that $s_1(\bar{c}_1) = \text{NI}$. The best response of the low cost player 1, however, would depend on player 2's strategy. If player 2 invests, player 1 prefers I; if player 2 plays NI then player 1 is better off playing NI as well.

Table 4.5 Bayesian investment game **2** between player 1 and player 2

Player 1\ player 2	Invest [y]	Not invest [$1 - y$]
High investment cost $\bar{c}_1$ [ρ]		
Invest	$\bar{c}_1$; c_2	$\bar{c}_1 + P_{21}L_1$; $P_{22}L_2$
Not invest	$P_{11}L_1$; $c_2 + P_{12}L_2$	$P_{11}L_1(1 - P_{21}) + P_{21}L_1(1 - P_{11})$; $P_{22}L_2(1 - P_{12}) + P_{12}L_2(1 - P_{22})$
Low investment cost $\underline{c}_1$ [$1 - \rho$]		
Invest [x]	$\underline{c}_1$; c_2	$\underline{c}_1 + P_{21}L_1$; $P_{22}L_2$
Not Invest [$1 - x$]	$P_{11}L_1$; $c_2 + P_{12}L_2$	$P_{11}L_1(1 - P_{21}) + P_{21}L_1(1 - P_{11})$; $P_{22}L_2(1 - P_{12}) + P_{12}L_2(1 - P_{22})$

Let the expected costs of low-cost player 1 for their strategy I and NI, respectively, as a function of player 2's mixed strategy y, be:

$$u_1(\mathrm{I}, y, \underline{c}_1) = \underline{c}_1 + L_1P_{21}(1 - y),$$

$$u_1(\mathrm{NI}, y, \underline{c}_1) = L_1P_{21}(1 - 2P_{11})(1 - y) + L_1P_{11}.$$

The low-cost player 1 weakly prefers to invest if:

$$y \geq \bar{y} := \frac{\underline{c}_1 - L_1P_{11}(1 - 2P_{21})}{2L_1P_{11}P_{21}}$$

Denoting by x the probability that the low-cost player 1 chooses Invest, we can write the low-cost player 1's mixed-strategy best-response correspondence $x^*(y)$ as follows:

$$x^*(y) = \begin{cases} \{1\}, & y \geq \bar{y}, \\ \{0\}, & y < \bar{y}. \end{cases}$$

The next step is to find the best-response correspondence of player 2. With probability ρ player 2 faces a high-cost player 1, which definitely chooses Not Invest. With probability $1 - \rho$ player 2 faces a low-cost player 1 which chooses Invest with probability x and Not Invest with probability $1 - x$. If player 2 chooses to invest then its expected cost is

$$u_2(\mathrm{I}, x) = c_2 + L_2P_{12} - (1 - \rho)xL_2P_{12}$$

If player 2 chooses NI then its expected cost is

$$u_2(\mathrm{I}, x) = L_2P_{12} + L_2P_{22}(1 - 2P_{12}) - x(1 - \rho)L_2P_{12}(1 - 2P_{22})$$

Player 2 weakly prefers to invest if

$$x \geq \bar{x} := \frac{c_2 - L_2P_{22}(1 - 2P_{12})}{2(1 - \rho)L_2P_{12}P_{22}}$$

Table 4.6 Example of Bayesian investment cost game between player 1 and player 2

Player 1\ player 2	Invest [y]	Not invest [1−y]
High investment cost $\bar{c}_1$ [ρ]		
Invest	(15, 19.8)	(20, 20)
Not invest	(10, 39.8)	(14.8, 39.2)
Low investment cost $\underline{c}_1$ [$1-\rho$]		
Invest [x]	(9.9, 19.8)	(14.9, 20)
Not invest [1 − x]	(10, 39.8)	(14.8, 39.2)

Hence, the mixed strategy best-response correspondence of player 2 is as follows:

$$y^*(x) = \begin{cases} \{1\}, & x \geq \bar{x}, \\ \{0\}, & x < \bar{x}. \end{cases}$$

To find the Bayesian-Nash equilibria we need to find intersections of the players' best response correspondences. In the considered game, player 1 is either a player of a Prisoner's Dilemma game with probability ρ or a player of a Stag Hunt game with probability $1-\rho$. Its true type is private information of player 1, while player 2 is a type of Stag Hunt player and this is assumed to be common knowledge (known by player 2 as well as by player 1).

To illustrate the Bayesian-Nash equilibria for the 'Bayesian game 2', we assume the following illustrative model parameters: $\bar{c}_1 = 15, \underline{c}_1 = 9.9, c_2 = 19.8, P_{21} = 0.01, P_{12} = 0.02, P_{11} = 0.02, P_{22} = 0.02; L_1 = 500, L_2 = 1,000$

Table 4.6 further elaborates and illustrates the hypothetical example.

If the probability that player 2 chooses to invest is larger or equal to $\bar{y} = \frac{1}{2}$ then the low-cost player 1 will weakly prefer to *Invest* as well:

$$x^*(y) = \begin{cases} \{1\}, & y \geq \frac{1}{2}, \\ \{0\}, & y < \frac{1}{2}. \end{cases}$$

Player 2 faces a high-cost player 1 with probability ρ and a low-cost player 1 with probability $1-\rho$. It weakly prefers to Invest if the probability that player 1 plays Invest is larger or equal to $\bar{x} := \frac{3}{4(1-\rho)}$, and thus the mixed strategy best-response correspondence of player 2 is as follows:

$$y^*(x) = \begin{cases} \{1\}, & x \geq \frac{3}{4(1-\rho)} \\ \{0\}, & x < \frac{3}{4(1-\rho)}. \end{cases}$$

The best response of player 2 thus depends on player 1 high-cost probability ρ. Notice that for $\rho \in [0,1]$, $\bar{x} \in \left[\frac{3}{4}, +\infty\right)$. Now we can find the Bayesian-Nash equilibria of the game determined by the high-cost probability ρ by finding the intersection of the players' best responses. Figures 4.1, 4.2, 4.3 plot graphs x^* and y^* for three cases:

1. $\rho \in \left[0, \frac{1}{4}\right)$
2. $\rho = \frac{1}{4}$
3. $\rho \in \left(\frac{1}{4}, 1\right]$

The Bayesian Nash equilibrium can be written in the form of a triple:

$$((\mathrm{NI}, x \circ \mathrm{I} \oplus (1-x) \circ \mathrm{NI}), y \circ \mathrm{I} \oplus (1-y) \circ \mathrm{NI})$$

The operators used in this equation are the following. The first operator ∘ is the Hadamar multiplication, while the second operator ⊕ indicates an exclusive disjunction. The meaning of the entire expression is presented hereafter.

We list the strategy of player 1 between brackets. The first element is the dominant action Not Invest of the high cost player 1, the second element is the mixed action of the low cost player 1. The mixed action of low cost player 1 is composed of the mixture of Invest with probability x and Not Invest with probability $(1 - x)$. The third element in the triple is the player 2's mixed strategy, which can be seen as the mix of Invest with probability y and Not Invest with probability $(1 - y)$.

Furthermore, an equilibrium is always indicated by a ★. Thus, in Figs. 4.1, 4.2, and 4.3, each equilibrium is marked with a ★.

Figure 4.1 represents the case in which player 1 is likely to have a low cost of investment, $\rho < \frac{1}{4}$. Here the character of the equilibria is determined by the low-investment cost game: there are two pure strategy equilibria ((NI, NI), NI), and ((NI, I), I), corresponding, respectively, to the low cost player 1 and player 2 playing both Not Invest in the first equilibrium, and playing both Invest in the second equilibrium. There is also a mixed strategy equilibrium, in which the low

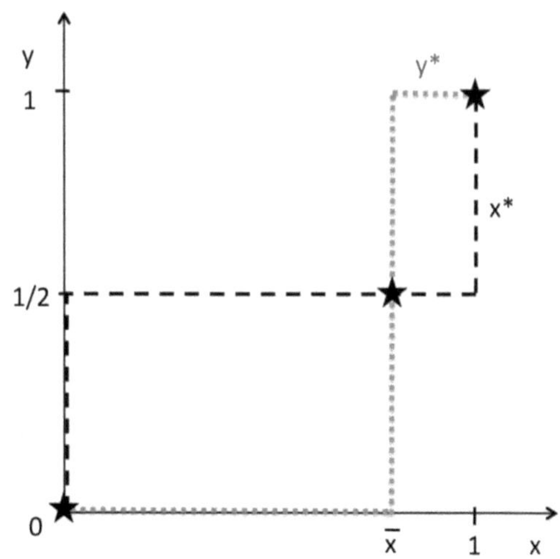

Fig. 4.1 Best-response correspondence for the low-cost player 1 and player 2, when the probability $\rho \in \left[0, \frac{1}{4}\right)$ that player 1 has a high cost of investment

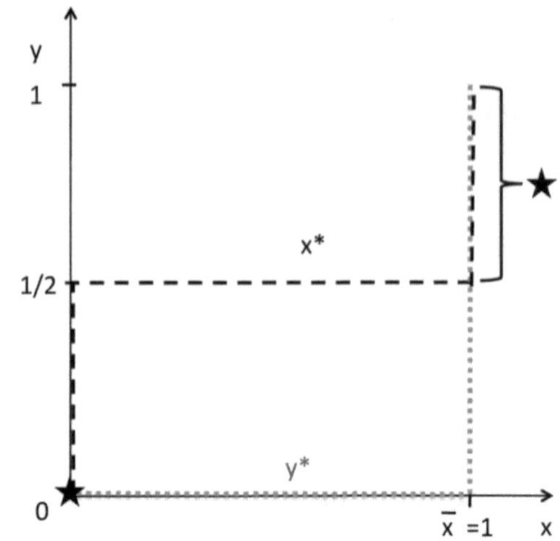

Fig. 4.2 Best-response correspondence for the low-cost player 1 and player 2, when the probability $\rho = \frac{1}{4}$ that player 1 has a high cost of investment

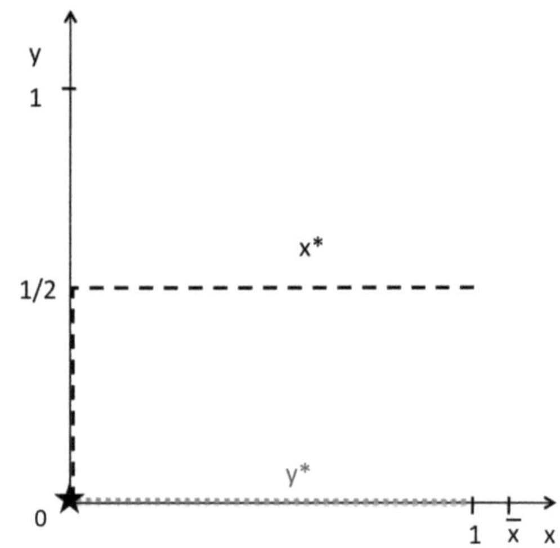

Fig. 4.3 Best-response correspondence for the low-cost player 1 and player 2, when the probability $\rho \in \left(\frac{1}{4}, 1\right]$ that player 1 has a high cost of investment

cost player 1 plays Invest half of its time and player 2 plays Invest with probability $\bar{x}$. As the probability $\rho \to \frac{1}{4}$ the first pure strategy Bayesian-Nash equilibrium remains the same, and the second pure equilibrium and the mixed equilibrium merge into a continuum of equilibria, as shown in Fig. 4.2.

When player 1 has a high investment cost with probability ρ larger than $\frac{1}{4}$ then the Bayesian game has a single pure strategy equilibrium in which both low-cost and high-cost types of player 1 and player 2 play Not Invest as Fig. 4.3 shows.

Figures 4.1 to 4.3 show how the solution changes when the probability ρ, which indicates that player 1 has a high cost of investment (or not), is changing. The method of funding Bayesian Nash equilibrium can straightforwardly be extended for the game with N > 2, however, a graphical representation will no longer be useful. It is important to notice that in each game Bayesian equilibria depend on information structure assumptions specific to the situation, which this game describes.

Reference

Reniers GLL, Dullaert W, Soudan K (2009) Domino effects within a chemical cluster: a game-theoretical modeling approach by using Nash-equilibrium. J Hazard Mater 16(1/3):289–293

Chapter 5
An Algorithm to Enhance Safety Collaboration Within Chemical Industrial Parks

5.1 Introduction

The objective of this chapter is to provide a game-theoretic analysis of strategic cooperation on safety within chemical industrial clusters, considering the Multi-Plant Council, which has been introduced in Chap. 2. Following the game-theoretic approach described in Chaps. 3 and 4, we include the Multi-Plant Council as a leading player in the game between chemical plants. Based on the analysis in Chap. 4, we provide insights into strategic incentives toward collaboration among the plants belonging to a chemical cluster.

As indicated in the first two chapters, it makes sense that, since several companies are involved in external domino effects, it is to the best interest of all plants composing a chemical industrial cluster to join forces in optimizing cross-plant loss prevention and making it as effective and as efficient as feasibly possible. By collaborating, costs can be minimized due to different types of collaboration benefits (joint investments, cut redundancies, joint training sessions, joint emergency exercises, etc.). Also, if companies decide to cooperate, safety can be optimized in a chemical cluster on an operational level (e.g. by exchanging information and data or implementing certain preventive measures), on a tactical level (e.g. by carrying out cross-plant risk assessments), as well as on a strategic level (e.g. make important long-term joint prevention and mitigation investments). However, research learns that companies are not inclined to cooperate on all these levels due to various reasons. Especially on a strategic level, firms are unwilling to cooperate due to trust and confidentiality concerns. Conceiving an easy-to-use and easy-to-understand approach for encouraging companies to collaborate as regards cross-plant prevention investments may thus be highly relevant toward more safety in an industrial area, and toward application of the ORDER Framework (cfr. Chap. 2) on a multi-plant level, leading to the next generation of chemical clusters.

Consider the game of chemical plants as introduced in Sect. 4.2. For the case where there are two Nash equilibria—either all companies invest in domino effects prevention or none of the companies do—the possibility of so-called 'tipping' exists, indicating that inducing some companies to invest in domino effects

G. Reniers and Y. Pavlova, *Using Game Theory to Improve Safety within Chemical Industrial Parks*, Springer Series in Reliability Engineering,
DOI: 10.1007/978-1-4471-5052-7_5, © Springer-Verlag London 2013

prevention will lead others to do so as well. If two Nash equilibria are involved, the socioeconomic optimal solution will be for every plant within the cluster to invest in domino effects prevention.

When there is only a single Nash equilibrium, the domino effects prevention investment choices made by every individual chemical facility will also be socioeconomically optimal in some situations. The case where the costs are sufficiently low so that each company wants to invest in domino effects prevention, even if all the other companies did not incur these costs, is a straightforward example of such a situation. If domino effects prevention investments would appear to be very high to every company relative to their potential benefits, then it might be efficient for no company to incur them.

In this chapter, we thus investigate the possibility of tipping the Nash equilibria from a state of 'no investment' to one of 'universal investment' (by all companies) as regards domino effects prevention in a chemical cluster.

5.2 The Domino Effects Game Model Revisited: N-Person Games

As we know from previous chapters, game theory is the theory of independent and interdependent decision making. Until now, we have mostly discussed games between two players for keeping the theory of making interdependent strategic decisions simple. However, in industrial practice, games usually involve more than two players. In game theory, such situations are dealt with by 'N-person games'. N-person games of strategy are games involving three or more players, each of whom has partial control over the outcome. It is obvious that such games potentially involve coalitions. The domino effect game discussed earlier in Sect. 4.2, can be seen as a partially cooperative N-person game in which coalition forming is allowed and even essential. Let a 'critical coalition' be a coalition of chemical companies where a change from 'not investing in domino effects prevention' to 'investing in domino effects prevention' by its members will induce all non-members of the coalition to follow suit.

Let us now revisit the domino effect game. Consider n chemical plants (players) composing a chemical cluster $\{\varphi\}$ and involved into conflict of investment decision. As previously in Chap. 4, every company i, $i = 1,\ldots,n$, is characterized by the probability $0 \le P_{\mathrm{ii}} \le 1$ that company i's lack of action can lead to an internally induced loss L_i, $L_i \ge 0$; the probability $0 \le P_{\mathrm{ji}} \le 1$ that company j's lack of action can cause an externally induced loss L_i to the company $i, j \ne i = 1,\ldots,n$; the investment $c_i \ge 0$ into cross-player prevention, which not only secures company i from occurrence of a major accident with escalation potential, but also guarantees no cross-border effect from i to any other company in the cluster; a discrete choice of strategy A_i that can take values Invest (I) or Not Invest (NI) into cross-player prevention of company i.

A loss caused by an internal domino effect is called ‘a direct loss’, whereas a loss to other companies (caused by an external domino effect) is considered as ‘an indirect impact’ and is referred to as ‘an indirect loss’. Let $l_i(\{\psi\}, A_i)$ be the expected indirect loss to a company i when it chooses a strategy A_i and ψ is a set of companies in the chemical cluster which choose strategy I($\psi \subseteq \phi$). Furthermore, if every other company than company i invests in domino effects prevention, then company i cannot suffer indirect impacts (i.e., impacts from other companies). In other words, if $\{\psi\} = \{1, 2, \ldots, i-1,\ i+1, \ldots, n\}$ then $l_i(\{\psi\}, A_i) = 0$, independent of the situation where player i invests or does not invest in domino effects prevention.

Consider a one stage game $\Gamma = \{A_1, \ldots, A_n;\ u_1, \ldots, u_n\}$, where u_i ($i = 1, \ldots, n$) is cost (negative payoff) for player/plant i. Cost for player i when it chooses strategy $A_i = \mathrm{I}$, given players from a coalition ψ do so as well, and the rest of the players $\phi \backslash \psi$ choose NI, is:

$$u_i^{\psi} = c_i + l_i(\{\psi\}, \mathrm{I})$$

If player i’s strategy is $A_i = \mathrm{NI}$, then player i’s cost is:

$$u_i^{\phi\backslash\psi} = L_i P_{\mathrm{ii}} \prod_{j \neq i,\, j \in \phi\backslash\psi} (1 - P_{\mathrm{ji}}) + l_i(\{\psi\}, \mathrm{NI})(1 - P_{\mathrm{ii}})$$

The first term is an expected direct loss of the company i, and the second term is its expected indirect loss. As already mentioned in the previous chapter, these conditions result from the fact that a chemical installation can only explode or be destroyed once or that internal and external domino effects do not originate at the same time (and since this book deals with safety (and not security), this assumption can be justified).

A chemical company i is indifferent between investing and not investing in domino effects prevention if:

$$c_i + \mathrm{l}_i(\{\psi\}, \mathrm{I}) = L_i P_{\mathrm{ii}} \prod_{j \neq i, j \in \{\varphi\backslash\psi\}} (1 - P_{\mathrm{ji}}) + l_i(\{\psi\}, \mathrm{NI})(1 - P_{\mathrm{ii}})$$

Hence, we can define the cost of investment in domino effects prevention at which company i is indifferent:

$$\tilde{c}_i(\{\psi\}) = L_i P_{\mathrm{ii}} \prod_{j \neq i, j \in \{\varphi\backslash\psi\}} (1 - P_{\mathrm{ji}}) + \mathrm{l}_i(\{\psi\}, \mathrm{NI})(1 - P_{\mathrm{ii}}) - \mathrm{l}_i(\{\psi\}, I)$$

If $c_i > \tilde{c}_i(\{\psi\})$, then company i will not invest in domino effects prevention, and vice versa.

Furthermore, if a chemical company within a chemical cluster φ decides to invest in domino effects prevention, the sub-cluster ψ is increased by one unit. As a result, the expected indirect loss, $P_{\mathrm{ji}} l_i(\{\psi\}, \mathrm{NI})$, decreases. Following, $\tilde{c}_i(\{\psi\})$ increases in ψ, since more chemical companies investing in domino effects prevention leads to lower expected indirect losses. Thus, the maximum cost at which domino effects prevention investments are justified, increases with every company deciding to invest in such prevention.

A Nash equilibrium for the multiperson domino effects game is a set of pure strategies $(A_1, \ldots, A_n)$ such that

1. $A_i = I, \forall i \in \psi$ (which may be empty),
2. if $\tilde{c}_i(\{\psi\}) > c_i$ then $A_i = \mathrm{I}$,
3. if $\tilde{c}_i(\{\psi\}) < c_i$ then $A_i = \mathrm{NI}$, and
4. if plant i is indifferent between I and NI, then $\tilde{c}_i(\{\psi\}) = c_i$.

Previously, it has been proven that Nash equilibria exist for this kind of game-theoretic problem [Heal and Kunreuther (2007) and Milgrom and Roberts (1990)]. There may be equilibria where all chemical companies invest in domino effects prevention, those where none do, and asymmetric pure strategy equilibria where some plants invest and others do not (see Proposition 4.3). If there are two equilibria, one with all enterprises not investing and the other with everyone investing in domino effects prevention, then it should be investigated in how to possibly tip the socially suboptimal equilibrium (NI, NI,…, NI) to a socially optimal equilibrium (I, I,…, I). Hence, we examine how the multi-person domino effects game with two (or more) equilibria may be tipped by a change in the strategy choices of a small number of players.

5.3 The 'Tipping Inducing Sub-Cluster' Concept

Let (NI, NI,…, NI) be a Nash equilibrium. A Tipping Inducing Sub-Cluster (TISC) for this equilibrium is a set ζ of chemical companies such that if $A_i = \mathrm{I}, \forall i \in \zeta$, then $\tilde{c}_j(\{\zeta\}) \geq c_j,\ \forall j \notin \zeta$. In other words, a TISC is a sub-cluster with the property that if all chemical plants belonging to that sub-cluster decide to invest in domino effects prevention, then for all other companies belonging to the entire chemical cluster the best strategy is also to invest in such prevention. A 'Minimum TISC' is a TISC of which no subset is also a TISC, indicating that all companies in the Minimum TISC are required to tip the other (non-investing) companies into a domino effects investment strategy. Furthermore, since there can be several Minimum TISCs and we are only interested in the one containing the smallest number of companies, we let the 'Smallest Number Minimum TISC' be a Minimum TISC where no other TISC includes fewer companies. Assume that the change in the expected indirect loss to plant i, which does not invest in domino effects prevention when plant j joins the set ψ of companies who have already invested in domino effects prevention, is:

$$l_i^j(\{\psi\}, \mathrm{NI}) = l_i(\{\psi \backslash j\}, \mathrm{NI}) - l_i(\{\psi\}, \mathrm{NI}) \geq 0, \forall j \in \psi$$

Heal and Kunreuther (2007) then prove (in general terms) that a Smallest Number Minimum TISC is easily characterized. They further indicate (again, in general terms) that an equilibrium where no company invests (e.g., in domino effects prevention) may be converted to one with full investment by persuading a subset of the companies to change their policies. Hence, the least expensive way of changing the equilibrium is providing incentives for a TISC to change its behavior, which will then tip the entire cluster. Hence, a TISC does exist in theory.

In order to help clear understanding, the next section discusses how such tipping might occur in a chemical cluster consisting of three companies. First, the discussion is kept in general terms and at the end, numerical values are used to show the N-person domino effects game's validity and its potential in a real industrial setting.

5.4 The TISC Concept Applied to a Cluster Composed of Three Plants

Consider a cluster of three companies 1–3. An investigation is conducted whether it is possible, and under what conditions, for one company by changing its strategy from NI to I, to tip the other two companies to change their strategies as well from NI to I.

Let us recapitulate some essential information mentioned before. The factor P_{ij} represents the probability that a domino event will occur in plant j caused by an event which takes place in plant i (in other words, P_{ij} is the likelihood of an external domino effect from company i to company j). If $i = j$, then the factor expresses the probability for an internal domino effect in company i. Every company can decide either to invest in domino effects prevention at a cost c_i (strategy I) or not to invest (strategy NI). If a domino effect takes place, the loss to company i equals L_i. For simplicity, domino effects prevention measures are assumed to be completely effective. Hence, if domino effects prevention investments are made in company i, no domino effect can originate from company i.

To investigate whether it is possible in the three-company case study for one company to tip the other two companies into changing their strategies, the existing Nash equilibria need to be determined, and the conditions under which strategies are dominant have to be established. Therefore, this section includes the cost matrices (i.e., negative payoff matrices) for two possible cases: (1) the strategy of company 1 is 'I', and (2) the strategy of company 1 is 'NI'.

The case where the strategy of company 1 is 'I' is first considered. If all three companies invest in domino prevention, then their costs are just their investment costs, c_i. If (besides company 1) company 2 invests in domino effects prevention, and company 3 does not invest, then companies 1 and 2 incur their investment cost c_i plus the expected loss from a domino effect from company 3 onto, respectively, company 1 or company 2 (i.e., $P_{31}L_1$, respectively $P_{32}L_2$). Company 3 just has an expected loss from an internal domino effect, i.e., $P_{33}L_3$. If neither company 2 and company 3 invest, then company 1 has an expected loss of its investment costs

Table 5.1 Cost matrix of companies 2 and 3 in the case where the strategy of company 1 is I

Company 2\ *Company* 3	Invest	Not invest
Invest	c_2; c_3	$c_2 + P_{32}L_2$; $P_{33}L_3$
Not Invest	$P_{22}L_2$; $c_3 + P_{23}L_3$	$P_{22}L_2(1 - P_{32}) + P_{32}L_2(1 - P_{22})$; $P_{33}L_3(1 - P_{23}) + P_{23}L_3(1 - P_{33})$

plus the expected loss from a domino effect from company 2 onto company 1 (i.e., $P_{21}L_1$), conditioned on there being no domino effect from company 3 onto company 1 (i.e., times $1 - P_{31}$), plus the expected loss from a domino effect from company 3 onto company 1 (i.e., $P_{31}L_1$), conditioned on there being no domino effect from company 2 onto company 1 (i.e., times $1 - P_{21}$). The conditions result from the fact that an installation can only explode once and that different domino effects are not initiated at the same time (since, as mentioned before, this book is about safety, and not security). All the other expected losses composing the cost matrix are determined in a similar way. The resulting cost matrix can be found in Table 5.1.

In case company 1's strategy is to invest (I), choosing I is a dominant strategy for company 2 under the conditions that

$$\begin{cases} c_2 < P_{22}L_2 \\ c_2 + P_{32}L_2 < P_{22}L_2(1 - P_{32}) + P_{32}L_2(1 - P_{22}) \end{cases}$$

Or:

$$\begin{cases} c_2 < P_{22}L_2 \\ c_2 < P_{22}L_2(1 - 2P_{32}) \end{cases}$$

The first condition is obviously what we would expect to be true in case of a single chemical company deciding whether to invest in domino effects prevention or not, thereby not taking into account the strategies of the other two companies within the cluster of three companies. The second condition, expressing the domino effect risk from a nearby company (thus considering the existence of the cluster in which the company is situated), is evidently stricter than the first condition.

Furthermore, (I, I) is a Nash equilibrium if $c_i < P_{ii}L_i$, and *Invest* is a dominant strategy for player $i = 2,3$ if $c_i < P_{ii}L_i(1 - 2P_{ji}), j \neq i = 2$ or 3. Situation (NI, NI) is Nash equilibrium if $c_i > P_{ii}L_i(1 - 2P_{ji})$, and *Not Invest* is a dominant strategy if $c_i > P_{ii}L_i$ (see Proposition 4.1 for correspondence between equilibriums and players' parameters in a 2-person game).

For the case where company 1 does not invest (strategy NI), a matrix representing the costs incurred by companies 2 and 3 may be determined as well. Table 5.2 illustrates this matrix.

In the case company 1's strategy is to not invest (NI), choosing I is a dominant strategy for company 2 under the conditions that

Table 5.2 Cost matrix of companies 2 and 3 in the case where the strategy of company 1 is NI

Company 2\ *Company* 3	Invest	Not invest
Invest	$c_2 + P_{12}L_2$; $c_3 + P_{13}L_3$	$c_2 + P_{12}L_2(1 - P_{32}) + P_{32}L_2(1 - P_{12})$; $P_{33}L_3(1 - P_{13}) + P_{13}L_3(1 - P_{33})$
Not invest	$P_{22}L_2(1 - P_{12}) + P_{12}L_2(1 - P_{22})$; $c_3 + P_{13}L_3(1 - P_{23}) + P_{23}L_3(1 - P_{13})$	$P_{22}L_2(1 - P_{12})(1 - P_{32}) + P_{22}L_2(1 - P_{22})(1 - P_{12}) + P_{22}L_2(1 - P_{12})(1 - P_{32})$ $P_{33}L_3(1 - P_{13})(1 - P_{23}) + P_{13}L_3(1 - P_{33})(1 - P_{23}) + P_{23}L_3(1 - P_{33})(1 - P_{13})$

$$\begin{cases} c_2 + P_{12}L_2 < P_{22}L_2(1 - P_{12}) + P_{12}L_2(1 - P_{22}) \\ c_2 + P_{12}L_2(1 - P_{32}) + P_{32}L_2(1 - P_{12}) < P_{22}L_2(1 - P_{12})(1 - P_{32}) + P_{12}L_2(1 - P_{22})(1 - P_{32}) \\ \qquad + P_{32}L_2(1 - P_{12})(1 - P_{22}) \end{cases}$$

The conditions for which I is a dominant strategy for company 3 can be derived analogously. Furthermore, (I, I) is a Nash equilibrium if $c_i + P_{ji}L_i < P_{ii}L_i(1 - P_{ji}) + P_{ji}L_i(1 - P_{ii})$ with $(i, j) = (2, 1)$ and $(i, j) = (3, 1)$.

Our N-person domino effects game is thus characterized by multiple Nash equilibria. Hence, if both the cost matrices from company 1 using 'I' (Table 5.1) or NI (Table 5.2) as a strategy are considered, the required conditions to turn the decision of companies 2 and 3 from 'not investing in domino prevention measures' (being part of a Nash equilibrium which is obviously not optimal from a socio-economic viewpoint) to 'investing in domino prevention measures' (being part of a socio-economic optimal Nash equilibrium) can be determined.

The tipping problem is illustrated using the following numerical example. Let $c_1 = 4{,}000$ €, $c_2 = c_3 = 700$ €. Assume that the probabilities aggregated for all installations within a company and per 10,000 years, are $P_{22} = P_{23} = P_{21} = P_{32} = P_{33} = P_{31} = 0.1$; $P_{11} = 0.2$; $P_{12} = P_{13} = 0.3$. Assume further that the potential company losses are $L_1 = 20{,}000$ € and $L_2 = L_3 = 10{,}000$ €.

It is examined what the dominant strategy is for company 1, given that companies 2 and 3 (considered in a sub-cluster of companies) are currently not investing in domino prevention ($\{\psi\} = \phi$) and do not consider the losses possibly resulting from the other companies in the cluster. In that case, the direct losses to companies 2 and 3 are $c_2(\phi) = c_3(\phi) = P_{22}L_2(1 - P_{12})(1 - P_{32}) = 600$€. It is obvious that, since $c_2 > c_2(\phi)$ and $c_3 > c_3(\phi)$, neither company 2 nor company 3 will invest in domino effects prevention if company 1's strategy is NI. Since $c_1(\phi) = P_{11}L_1(1 - P_{31})(1 - P_{21})$ =3,200 € is smaller than c_1 (=4,000 €), company *1* will indeed not invest and will indeed not invest and (NI, NI, NI) is actually a Nash equilibrium. Hence, if company 1 does not invest, then not investing is a dominant strategy for the other companies for all $c_i > 600$ €.

Given that the three companies are located within the same cluster, limiting the analysis to direct (internal) domino effects does not offer a solid basis for domino prevention management. Also the indirect loss caused by the fact that only a subset $\{\psi\}$ of companies is investing in prevention should be considered. An example of how including the indirect risks can alter the analysis, is therefore provided. Assume that company 1 is obliged to invest in domino effects prevention (e.g. due to national or regional regulations and/or legislation) and as a result no negative externality from company 1 is imposed on the other companies (2 and 3). If this is the case, the indirect risk to chemical company *i* (2 or 3) is the same whether company *i* (2 or 3) itself decides to 'invest' or decides to 'not invest', i.e., $l_i(\{\psi\}, \mathrm{NI}) = l_i(\{\psi\}, I)$ (where $i = 2, 3$), hence we can use expression

$$\tilde{c}_i(\{\psi\}) = P_i \cdot \left(L_i \prod_{j \neq i, j \in \{\varphi/\psi\}} (1 - P_j) - l_i(\{\psi\}, \text{NI}) \right)$$

to determine the critical cost levels of companies 2 and 3, which in both cases amount to 800 €.

Since both c_2 and c_3 are smaller, companies 2 and 3 will be changing their strategy of NI to I as a result of company 1 doing so. Therefore, company 1 has the power to tip the other companies within the three-company cluster from one strategy to another strategy. In other words, one company's strategic choice concerning domino prevention may significantly influence the other companies' domino prevention-related strategic decisions in a chemical cluster. This is an important finding implying that e.g., company-specific incentives or well-elaborated domino prevention regulations can lead to substantial safety improvements within chemical clusters.

5.5 Searching the TISC for Chemical Clusters Composed of 3 Chemical Plants or More

Real-life chemical industrial clusters obviously often consist of more than three plants situated in each other's neighborhood. Hence, the actual situation in real cluster cases is much more complex to deal with. We investigate in this section whether it is possible to take more complicated situations into account by establishing the potential role of cluster safety governance. We stress that cluster safety governance can actually be realized and implemented through self-regulation by a multi-plant council (MPC), as explained in Chap. 2. In Sect. 2.2, we suggested the setting up of an institution at the multi-plant-level, the so-called MPC, which would be responsible for a continuous follow-up of external safety (and security) improvements at the individual companies belonging to the industrial multi-plant cluster. Due to its cross-plant trust inducing capability, the Multi-Plant Council might play a stimulating role to reach the socioeconomic optimum. The MPC's responsibilities and structures exceed those of any existing collaborative bodies. Nonetheless, the MPC's independent experts (cfr. Sect. 2.2) might be able to use the collected confidential information in the individual plants as input for executing computer-automated software, audits, inspections, etc., and also for playing a 'sequential-move domino effects game', as explained further in this section.

In the next Sect. 5.5.1, a game-theoretic interpretation of a conflict between safety in a chemical cluster and social cooperation is described. A two-stage sequential-move game between the chemical plants and the MPC is introduced in Fig. 5.1. In the considered game, the MPC is a leader player who has the opportunity to decide to support or not cooperation among the cluster companies. The MPC's objective is to achieve full cooperation among players through establishing a system of incentives (e.g., subsidy system) at minimum expense.

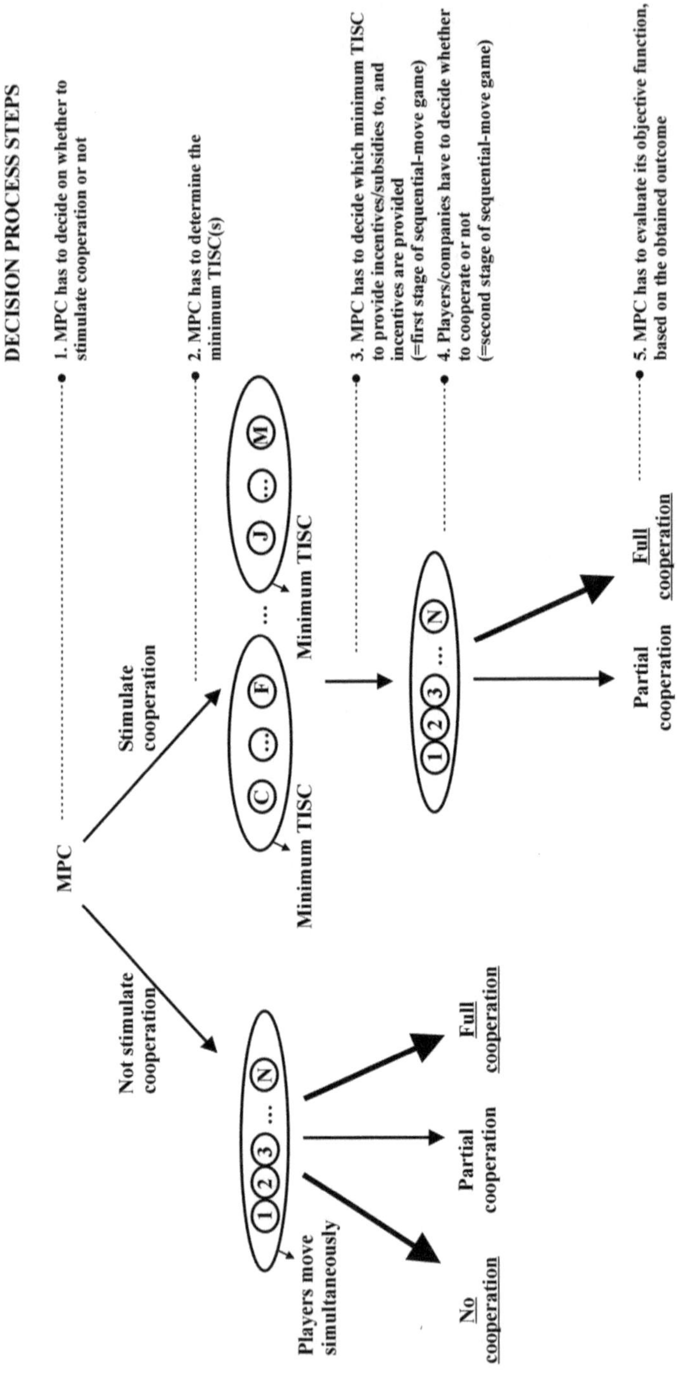

Fig. 5.1 Extensive form of the two-stage sequential-move game

The individual chemical companies are followers. After the leader makes the move, the followers may decide to invest in cooperative prevention of domino accidents and play Nash equilibrium. The solution of the game is obtained as a subgame-perfect equilibrium. To establish an optimal system of incentives, we search for a minimal coalition of players, whose initial decision to cooperate is sufficient to induce cooperation among the rest of the players, i.e., a TISC. Section 5.5.2 contains an algorithm of identifying the minimal TISC and the roadmap for cooperation enhancement. The algorithm is presented in Fig. 5.2 and then explained in detail. Section 5.6 contains an illustrative example, which describes a game between five heterogeneous companies (forming the considered multi-plant cluster) and the MPC. The purpose of the example is to demonstrate how such a stepwise plan given in Sect. 5.5.2 to improve cross-company safety management in a chemical industrial cluster, can be implemented.

5.5.1 Two-Stage Sequential-Move Game of Domino Effect Prevention

Let us consider a two-stage sequential-move game between $n + 1$ players: n chemical plants and the MPC. Sequential-move games and solution methods were previously discussed in Sect. 3.4 *Extensive form games with discrete set of strategies*. The MPC is considered as a leader player who makes the first move, and the individual companies are followers. The MPC's objective is to achieve full cooperation among the players (i.e., chemical companies composing the industrial multi-plant area) through establishing a subsidy system at minimum expense. The rest of the players rationally react to the subsidies proposed by the leader and simultaneously play Nash equilibrium.

As mentioned before, if a player i has decided to invest in domino effect prevention and a player j has to decide whether or not to do likewise, then the higher the probability that j can benefit of preventive investment by i, the less likely it is that j will follow suit and invest as well. The domino effect game considered in this section is a game with multiple Nash equilibria.[1] The cost structure of domino prevention implies that a player i's individual cost of preventive investment is lower than the damage to player i in the case that an internal domino accident would happen. In such a situation, it is obvious that if company i acts as a rational player, it is assumed to invest. All other players will follow suit. Additionally, if no player invests then player i does not have incentives to invest (alone). In such a case of conflict between safety and social cooperation among chemical companies within a chemical multi-plant area (or cluster under consideration), the game can be interpreted as a 'coordination game with assurance' (Skyrms 2004), with full cooperation and non-cooperation as two Nash equilibria.

[1] See Proposition 4.3.

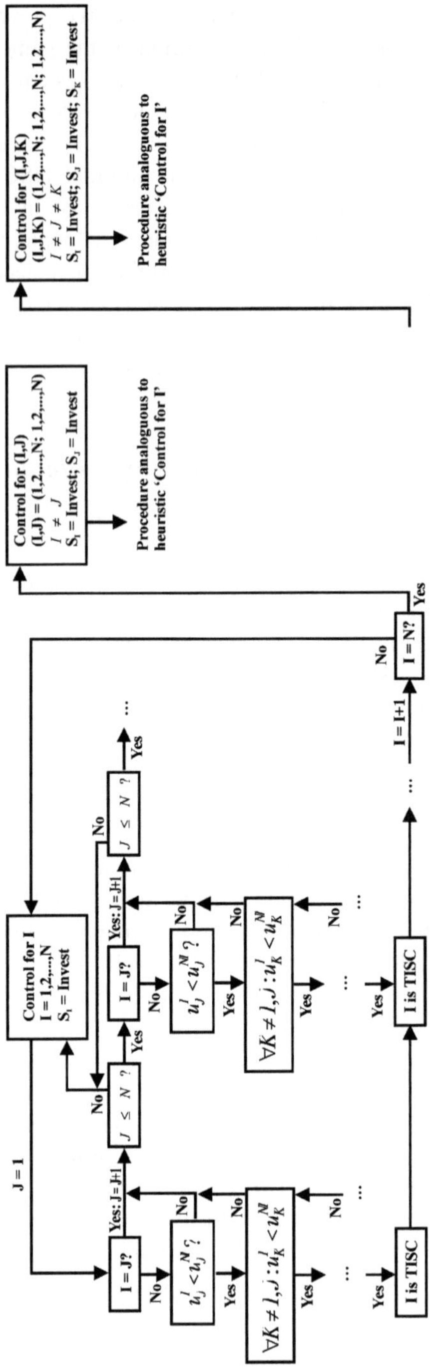

Fig. 5.2 Algorithm for determining all TISCs within a chemical industrial area

Thus, investment in preventive measures can be considered as a public good and its provision is conditioned on whether other players also invest, or do not invest.

The MPC aims at enhancing cooperation among the chemical plants of the chemical multi-plant area by providing them with subsidies at minimum expense. The rest of the players (plants), who do not obtain any subsidies, are assumed to react rationally to the decision of the MPC and simultaneously play Nash equilibrium. To provide game theory based guidance for the MPC, we are interested in a subgame-perfect equilibrium in such an N-person game.

As already explained in Sect. 3.4.2, the subgame-perfect equilibrium may be found by backward induction, an iterative process for solving finite extensive form games or sequential games. First, the optimal strategy of the player who makes the last move of the game is determined. Second, the optimal action of the next-to-last moving player is determined taking the last player's action as given. The process continues in this way backwards in time until all players' actions have been determined.

Because simultaneous-move games of coordination can have two possible equilibrium situations, a leader player can direct the outcome toward the one which is more preferable to its payoff. Heal and Kunreuther (2007) and Reniers et al. (2009) both give an insight into the potential role of a mediator in case of these types of games. In the sequential-move game of domino accident prevention, the objective of the leader player (MPC) is explicitly formulated as follows:

$$u_0 = -\sum_{i \in \psi} c_i + \sum_{i \in \phi} \left(\mathrm{u}_i^{\mathrm{NC}} - \mathrm{u}_i \right). \tag{5.1}$$

The objective function for the MPC is composed of two parts: the first term represents the MPC's costs of providing cooperation incentives to the sub-group ψ of players, whereas the second term represents the MPC's benefits from players cooperating. The first term can be explained as follows. Due to the extremely low probabilities of a domino effect occurring, players are not inclined to invest in preventive measures. The players have to be provided incentives which are equivalent (in monetary terms) to the expected costs of investment. Hence, the sum of the prevention investment costs c_i for all players in a TISC is the amount of subsidies required to tip these companies from a strategy *Not Invest* to a strategy *Invest*.

The second term describes the difference between expected costs of the players in the situation, which would be realized at the end of the game (u_i), and the losses in the undesirable situation when nobody cooperates (u_i^{NC}). Given that the MPC's objective is to achieve full cooperation among players, its payoff can be justified since the MPC's benefits become larger if more players cooperate, and they are largest in the case that all players cooperate.

Now let us describe the extensive form of the game (see Fig. 5.1). In the terminal nodes, the players and the MPC receive payoffs according to formulae (4.1–4.2) and (5.1)

As illustrated in Fig. 5.1, the MPC has two strategies: (1) to 'stimulate cooperation', and (2) to 'not stimulate cooperation'. If the MPC chooses not to enhance collaboration within the multi-plant area, the individual companies/players play a Nash equilibrium and the real situation may either lead to no cooperation (and no joint investment) or to full cooperation (joint investments). In current industrial practice (where no organization such as the MPC exists), the former situation exists in chemical industrial clusters. In the case the MPC chooses to invest and to stimulate collaboration within the cluster, a subset of players will be given incentives/subsidies on the condition that they cooperate. As explained before, such a subset of players should be a minimum Tipping Inducing Sub-Cluster (minimum TISC).

Definition 5.1. A **TISC** is a coalition with the property that if all of its members have a strategy *Invest*, then for all other individual players of the greater cluster where the TISC is part of, the best response is also to invest.

Definition 5.2. A **minimum TISC** is a TISC of which no subset is also a TISC.

According to this concept, the MPC can identify a group of players (or all groups of players), whose initial willingness to cooperate allows to sustain full cooperation. This observation suggests that an algorithm for determining a TISC in a sequential-move game is one of the important steps to solving it.

5.5.2 *TISC-Algorithm and Roadmap for Safety Cooperation Enhancement Within Chemical Industrial Clusters*

In this subsection, we present an algorithm that can be used by the MPC for determining TISCs. We consider a subgame, which corresponds to the MPC's choice of strategy 'Stimulate collaboration'. In this subgame, first a TISC composed of players from ψ chooses to play 'Invest' and then the remaining $\varphi\backslash\psi$ group of players sequentially decides to cooperate or not to cooperate. If we assume that all TISCs have been successfully identified then the bold lines on the second stage in Fig. 5.1 describe Nash equilibria in each of two sub-games. Without loss of consistence we assume that in the second sub-game (when the MPC plays 'Not to stimulate cooperation') the players move simultaneously. It implies that none of the players has a right or commitment to decide before others and thus the coordination game has two possible outcomes: no cooperation and full cooperation. Later in this subsection, we discuss that without coordination in a form of a system of incentives, the players are not willing to switch from strategy 'Not invest' to 'Invest'.

An algorithm was developed to determine all TISCs within a chemical industrial cluster. The algorithm is provided in Fig. 5.2.

The TISC-algorithm first verifies whether every company, as a single company, may act as a TISC. Suppose a player I is a TISC. To prove that I is a minimal TISC indeed, we need to show that if the player I plays 'Invest' then we can find another player *J*, other than I, who prefers playing 'Invest' to 'Not Invest'. If such a player J exists, then we need to check if there is a player *K* (other than I or *J*), who prefers to play 'Invest' to 'Not Invest' if the players I and J play 'Invest'. The routine continues until all players from φ play 'Invest'. If such a choice of player I, being 'Invest', can induce other players sequentially playing Invest as well, then player I is a TISC. Otherwise, if there is no such a single player, we need to check all couples of players to be TISCs or not. In a similar way, we should assume that some players I and J from φ play 'Invest'. Then we need to check if there is a player *K* other than I or *J* who will prefer 'Invest'. Continuing in a similar way, all triples of companies are to be checked, etc. This process is repeated until all $(n-1)$-multiples are validated against having the potential of being a Tipping Inducing Sub-Cluster. At the end of the algorithm, all possible TISCs within a chemical industrial cluster are identified. It is necessary to point out that a TISC always exists because the static game is a coordination game with assurance where both noncooperative and cooperative outcomes are stable.

The next step is to determine the minimum TISCs, i.e., the TISCs of which no subsets are also TISCs. This stage can be carried out by simply comparing all TISCs one-by-one. If all minimum TISCs are determined, all information is available to the MPC to take incentive/subsidy decisions.

5.6 Stimulating Safety Collaboration in a Chemical Industrial Park: An Illustrative Example

A stepwise plan or Roadmap for stimulating safety cooperation can be set up. The Roadmap consists of the following six steps:

1. Fix the number of chemical companies where the MPC would like to enhance collaboration in-between.
2. Collect the required parameters from the companies to carry out the collaboration-enhancement study (i.e., potential losses, domino prevention costs, and domino accident probabilities).
3. Use the TISC-algorithm to determine all minimum TISCs.
4. Use the objective function of the MPC to identify those minimum TISCs that deliver the MPC optimal benefits.
5. Provide incentives/subsidies to the optimal TISC settled in the previous stage.
6. Once the game is played, evaluate the outcome and act accordingly.

These six steps can be used as a guide for an MPC to actively enhance collaboration between chemical companies in the field of domino prevention investments. After evaluation of the outcome, the MPC may decide to provide incentives to a minimum TISC, which delivers the largest (positive) payoff to the

MPC. Such a strategy is optimal for the MPC, since no other TISC containing the minimum TISC can increase the MPC's payoff. Indeed, expression (5.1) tells us that including more companies other than those belonging to the minimum TISC can only increase costs of incentives but will have no benefit effects from companies' cooperation. The complexity of the proposed TISC-algorithm increases combinatorial when the number of plants increases. The number of all possible combinations, which have to be tested in the TISC-algorithm, is $2^n - 2$ (with n the number of chemical companies). Given that in practice, n is a rather limited number, the proposed algorithm can be implemented.

To provide the reader with an idea of the usefulness of the algorithm, an illustrative example of a chemical multi-plant area is given hereafter. Let us assume that the industrial area is composed of five companies $i = \{1,2,3,4,5\}$, characterized with the following estimates of the players:

$$P_{5\times5} = \begin{pmatrix} 1.1\cdot10^{-4} & 0.055\cdot10^{-4} & 0.055\cdot10^{-4} & 0.055\cdot10^{-4} & 0.055\cdot10^{-4} \\ 0.055\cdot10^{-4} & 1.1\cdot10^{-4} & 0.055\cdot10^{-4} & 0.055\cdot10^{-4} & 0.055\cdot10^{-4} \\ 0.043\cdot10^{-4} & 0.043\cdot10^{-4} & 0.64\cdot10^{-4} & 0.032\cdot10^{-4} & 0.04\cdot10^{-4} \\ 0.043\cdot10^{-4} & 0.043\cdot10^{-4} & 0.032\cdot10^{-4} & 0.64\cdot10^{-4} & 0.04\cdot10^{-4} \\ 0.025\cdot10^{-4} & 0.025\cdot10^{-4} & 0.02\cdot10^{-4} & 0.02\cdot10^{-4} & 0.8\cdot10^{-4} \end{pmatrix};$$

$$\begin{aligned} c &= \left(1.3\cdot10^4, 1.3\cdot10^4, 10^4, 10^4, 0.78\cdot10^4\right); \\ L &= \left(1.6\cdot10^8, 1.6\cdot10^8, 10^8, 10^8, 10^8\right). \end{aligned}$$

Here, we assume that both costs of domino effect prevention as well as possible losses are given in euro (for example).

Furthermore, we assume that players 1 and 2 on the one hand, and players 3 and 4 on the other hand, have similar probabilities, costs, and losses. We now solve the two-stage sequential game backward. First, 30 possible situations in the simultaneous-move game of 'not stimulate cooperation' are feasible. Here players' costs are given as follows:

$$\begin{aligned}
u_i^{\psi}(\mathrm{I}_i, \mathrm{I}_j, I_k, \mathrm{I}_l, \mathrm{I}_m) &= c_i; \\
u_i^{\psi}(\mathrm{I}_i, \mathrm{I}_j, \mathrm{I}_k, \mathrm{I}_l, \mathrm{NI}_m) &= c_i + P_{\mathrm{mi}}L_i; \\
u_i^{\psi}(\mathrm{I}_i, \mathrm{I}_j, I_k, \mathrm{NI}_l, \mathrm{NI}_m) &= c_i + (P_{\mathrm{li}}(1-P_{\mathrm{mi}}) + P_{\mathrm{mi}}(1-P_{\mathrm{li}}))L_i; \\
u_i^{\psi}(\mathrm{I}_i, \mathrm{I}_j, \mathrm{NI}_k, \mathrm{NI}_l, \mathrm{NI}_m) &= c_i + (P_{\mathrm{li}}(1-P_{\mathrm{mi}})(1-P_{\mathrm{ki}}) + P_{\mathrm{ki}}(1-P_{\mathrm{li}})(1-P_{\mathrm{mi}}) + P_{\mathrm{mi}}(1-P_{\mathrm{li}})(1-P_{\mathrm{ki}}))L_i; \\
u_i^{\psi}(\mathrm{I}_i, \mathrm{NI}_j, \mathrm{NI}_k, \mathrm{NI}_l, \mathrm{NI}_m) &= c_i + \left(P_{\mathrm{li}}(1-P_{\mathrm{mi}})(1-P_{\mathrm{ki}})\left(1-P_{\mathrm{ji}}\right) + P_{\mathrm{ki}}(1-P_{\mathrm{li}})(1-P_{\mathrm{mi}})\left(1-P_{\mathrm{ji}}\right)\right)L_i \\
&\quad + P_{\mathrm{li}}(1-P_{\mathrm{mi}})(1-P_{\mathrm{ki}})\left(1-P_{\mathrm{ji}}\right)L_i; \\
u_i^{\varphi}(\mathrm{NI}_i, \mathrm{NI}_j, \mathrm{NI}_k, \mathrm{NI}_l, \mathrm{NI}_m) &= \left(P_{\mathrm{li}}(1-P_{\mathrm{mi}})(1-P_{\mathrm{ki}})\left(1-P_{\mathrm{ji}}\right)(1-P_{\mathrm{ii}}) + P_{\mathrm{ki}}(1-P_{\mathrm{li}})(1-P_{\mathrm{mi}})\left(1-P_{\mathrm{ji}}\right)(1-P_{\mathrm{ii}})\right)L_i \\
&\quad + \left(P_{\mathrm{mi}}(1-P_{\mathrm{li}})(1-P_{\mathrm{ki}})\left(1-P_{\mathrm{ji}}\right)(1-P_{\mathrm{ii}}) + P_{\mathrm{ji}}(1-P_{\mathrm{li}})(1-P_{\mathrm{ki}})(1-P_{\mathrm{mi}})(1-P_{\mathrm{ii}})\right. \\
&\quad \left.+ P_{\mathrm{ii}}(1-P_{\mathrm{li}})(1-P_{\mathrm{ki}})(1-P_{\mathrm{mi}})\left(1-P_{\mathrm{ji}}\right)\right)L_i; \\
u_i^{\varphi/\psi}(\mathrm{NI}_i, \mathrm{NI}_j, \mathrm{NI}_k, \mathrm{NI}_l, \mathrm{I}_m) &= \left(P_{\mathrm{li}}(1-P_{\mathrm{ki}})\left(1-P_{\mathrm{ji}}\right)(1-P_{\mathrm{ii}}) + P_{\mathrm{ii}}(1-P_{\mathrm{li}})\left(1-P_{\mathrm{ji}}\right)(1-P_{\mathrm{ki}})\right)L_i \\
&\quad + \left(P_{\mathrm{ji}}(1-P_{\mathrm{ki}})(1-P_{\mathrm{li}})(1-P_{\mathrm{ii}}) + P_{ki}(1-P_{li})\left(1-P_{ji}\right)(1-P_{ii})\right)L_i; \\
u_i^{\varphi/\psi}(\mathrm{NI}_i, \mathrm{NI}_j, \mathrm{NI}_k, \mathrm{I}_l, \mathrm{I}_m) &= \left(P_{\mathrm{ki}}\left(1-P_{ji}\right)(1-P_{\mathrm{ii}}) + P_{\mathrm{ji}}(1-P_{\mathrm{ki}})(1-P_{\mathrm{ii}}) + P_{\mathrm{ii}}\left(1-P_{\mathrm{ji}}\right)(1-P_{\mathrm{ki}})\right); \\
u_i^{\varphi/\psi}(\mathrm{NI}_i, \mathrm{NI}_j, \mathrm{I}_k, \mathrm{I}_l, \mathrm{I}_m) &= \left(P_{\mathrm{ji}}(1-P_{\mathrm{li}}) + P_{\mathrm{ii}}\left(1-P_{\mathrm{ji}}\right)\right)L_i; \\
u_i^{\varphi/\psi}(\mathrm{NI}_i, \mathrm{I}_j, \mathrm{I}_k, \mathrm{I}_l, \mathrm{I}_m) &= P_{\mathrm{ii}}L_i.
\end{aligned}$$

One should always keep in mind that these are the players' costs. Hence, the lower such a cost is, the better off the player is.

This coordination with assurance game has two equilibria:

1. _Pareto-Efficient Nash equilibrium:_ all players choose cooperation and they thus choose to invest in domino effects precautions:

$$u_1^{\psi}(\mathrm{I}_1,\mathrm{I}_2,\mathrm{I}_3,\mathrm{I}_4,\mathrm{I}_5) = u_2^{\psi}(\mathrm{I}_1,\mathrm{I}_2,\mathrm{I}_3,\mathrm{I}_4,\mathrm{I}_5) = 13{,}000;$$
$$u_3^{\psi}(\mathrm{I}_1,\mathrm{I}_2,\mathrm{I}_3,\mathrm{I}_4,\mathrm{I}_5) = u_4^{\psi}(\mathrm{I}_1,\mathrm{I}_2,\mathrm{I}_3,\mathrm{I}_4,\mathrm{I}_5) = 10{,}000;$$
$$u_5^{\psi}(\mathrm{I}_1,\mathrm{I}_2,\mathrm{I}_3,\mathrm{I}_4,\mathrm{I}_5) = 7{,}800;$$

2. _Risk-dominant Pareto Inefficient Nash equilibrium:_ each player chooses not to cooperate and thus chooses not to invest in domino effects preventive measures:

$$u_1^{\varphi}(\mathrm{NI}_1,\mathrm{NI}_2,\mathrm{NI}_3,\mathrm{NI}_4,\mathrm{NI}_5) = u_2^{\varphi}(\mathrm{NI}_1,\mathrm{NI}_2,\mathrm{NI}_3,\mathrm{NI}_4,\mathrm{NI}_5) = 20{,}255.383;$$
$$u_3^{\varphi}(\mathrm{NI}_1,\mathrm{NI}_2,\mathrm{NI}_3,\mathrm{NI}_4,\mathrm{NI}_5) = u_4^{\varphi}(\mathrm{NI}_1,\mathrm{NI}_2,\mathrm{NI}_3,\mathrm{NI}_4,\mathrm{NI}_5) = 13{,}869.336;$$
$$u_5^{\varphi}(\mathrm{NI}_1,\mathrm{NI}_2,\mathrm{NI}_3,\mathrm{NI}_4,\mathrm{NI}_5) = 9{,}899.669.$$

It is realistic to achieve full cooperation and indeed full cooperation is more preferable as it will bring a Pareto-Efficient outcome if all players act rationally (i.e., there is no other situation, which is more beneficial for all players). However, if one (or more) of the players fail(s) to collaborate, the other cooperating players will lose. Playing the 'inefficient' Nash equilibrium (NI,…,NI) is thus less risky for the players as the costs' variance over the other players' strategies is lower. Specifically, the Nash equilibrium, when all players choose to *Invest*, is Pareto optimal, while the other, when all players choose to *Not Invest*, is risk-dominant. Though the situation when all players cooperate is thus actually possible, observation of current industrial practice proves that the cooperative strategy is not credible and the situation when all players play the Pareto-Efficient Nash equilibrium is highly unlikely.

The MPC knows that if it does not stimulate collaboration, all players play *Not Invest*, and thus the MPC's payoff is zero (there are no incentive/subsidies costs and also no benefits ($u_i = u_i^{\mathrm{NC}}$) as no cooperation among players takes place).

Let us now consider an option when the MPC chooses to provide a group of companies with subsidies to switch from strategy '*Not Invest*' to strategy '*Invest*'. The MPC identifies those players whose initial willingness to cooperate is required to make the rest of the players follow suit (i.e., the minimum TISCs). There may be one or several minimum TISCs due to the fact that the players of the game are heterogeneous. Looking at the payoff structure of the players in the simultaneous-move game without any cooperation stimulation, by using the TISC-determination algorithm we notice that if players (3, 4) choose strategy *Invest*, then the rest of the players will follow suit.

Let us consider subset $\zeta = \{3,4\}$: we notice that then for players 1 and 2 Invest becomes a dominant strategy:

Once players 1 and 2 join ζ then player 5 also prefers an option *Invest*

It is easy to check that there is no other subset of players, which could have the properties of a TISC. Additionally, subset ζ is also a minimum TISC. According to formula (5.1), the payoff of the MPC is $u_0 = 20,000 - 30,800 + 44,024.388 = 33,224.388$.

Following backward induction, the subgame-perfect equilibrium of the game is when the MPC plays 'Stimulate cooperation'. In that case, subset $\zeta = \{3,4\}$ will choose to cooperate ('Invest in domino effects prevention'), and the rest of the players simultaneously or sequentially will choose to cooperate as well. Another candidate for sub-game perfect equilibrium (when the MPC plays 'Not stimulate cooperation' and the players play 'cooperate') can be eliminated because the MPC is informed that if it does not 'Stimulate cooperation', the only credible move of the players is to choose not to cooperate.

In this case, the MPC will decide to provide incentives only to the players from the minimum TISC $\zeta = \{3,4\}$, since the MPC's (positive) payoff cannot increase if incentives are provided for a wider subset of players. Such a decision is intuitively clear: From the matrix of probabilities (P), one may notice that the likelihood that the companies 3 and 4 will experience the domino accident from other companies, are the smallest, while the costs of prevention are still rather high (compared with the other companies). Thus, companies 3 and 4 are least vulnerable in the system and may not be inclined to start cooperation without additional incentives, even though full cooperation is more preferable than non-cooperation.

References

Heal G, Kunreuther H (2007) Modeling interdependent risks. Risk Anal 27(3):621–634

Milgrom P, Roberts J (1990) Rationalizability, learning and equilibrium in games with strategic complementarities. Econometrica 58:1255–1277

Reniers GLL, Dullaert W, Soudan K (2009) Domino effects within a chemical cluster: a game-theoretical modeling approach by using nash-equilibrium. J Hazard Mater 16(1/3):289–293

Skyrms B (2004) The Stag Hunt and Evolution of Social Structure. Cambridge University Press, Cambridge

Chapter 6
Cooperative Incentives Approach in Case of an MPC with Limited Resources

6.1 Introduction

The primary objective of the approach presented in this chapter is to explain how to achieve the best possible form of cooperation among the chemical plants in the cluster. In current industrial practice, this is done by means of exchange of confidential information, without possibility of subsidies and incentives. Such situations indeed occur when the Multi-plant council (MPC) has only limited resources available and financial incentives are not possible.

In line with the earlier presented game-theoretic analysis in Chaps. 4 and 5, we distinguish three types of cooperation regarding investment in domino prevention measures. First, a multi-plant area composed of chemical plants can fully cooperate, i.e., all chemical plants belonging to the area decide to invest in domino effect prevention measures. The second type is a partial investment case which occurs when at least one but not all, chemical plants belonging to the industrial area decide(s) to invest in domino effects precaution. Third, there is no investment at all regarding escalation precaution in a chemical industrial park.

The goal of this chapter is to develop a method for stimulating cooperation between two or more competing companies, leading to cross-plant precaution investments. As also explained previously, due to the interdependence of plants' strategy choices, collaboration benefits can be realized by gathering cross-plant information, processing it, and based on the calculations financially stimulating domino precaution cooperation/investments within a chemical industrial area.

In current industrial practice, chemical plants in a cluster are not inclined to cooperate regarding domino effect prevention measures, due to trust and confidentiality issues. Nonetheless, internal, and especially external domino effects, can only be minimized to the full extent through collaboration between adjacent companies. Therefore, it is important that our Cooperative Incentives Approach (CIA) should facilitate and stimulate cooperation, thereby taking into account information from different companies at once and considering different strategic collaboration options, in order to advise on optimized collaboration benefits.

G. Reniers and Y. Pavlova, *Using Game Theory to Improve Safety within Chemical Industrial Parks*, Springer Series in Reliability Engineering,
DOI: 10.1007/978-1-4471-5052-7_6, © Springer-Verlag London 2013

As before, expected losses from a domino accident are given as the product of the probability of a domino event over a certain time interval (e.g., a year) and the internally/externally induced losses expressed in financial units (Euro). Hence, domino accident expected damage, and also domino prevention investment costs, are expressed in Euro per year. Since the installations' lifetimes are thus factors to be taken into account when making a domino effect prevention decision, we assume that all chemical installations to be protected have still a minimum lifetime of at least one year (that is, at the time of making an *invest-or-not* decision). This is a reasonable assumption, since chemical installations, storage tanks, etc. in general are designed to have an average lifespan of at least several years, up to 10 years and sometimes much more, depending on e.g., maintenance and other factors.

If a chemical company within a chemical cluster decides to invest in domino effects prevention, the expected indirect loss within the industrial park decreases. Each non-investing plant adds an additional risk of domino effects from a catastrophic event in the cluster, and therefore additional externally induced losses for the other plants situated in the cluster. This implies that collaboration benefits (cfr. Sect. 5.1) increase as the number of investing plants increases. A maximization of collaboration benefits, however, does not necessarily entail that the total expected losses (expected investment costs plus expected damage from a domino accident) of the individual plants is minimized.

6.2 The Cooperative Incentives Approach

The CIA should analyze under what conditions and what type of cooperation is beneficial (in financial terms), as well as give insights into how such potential cooperation may be stimulated. In order to establish such an approach, we distinguish between various responsibilities of the stakeholders. First, the number of chemical plants in the multi-plant area needs to be determined. Second, the plant representatives (e.g. prevention managers, financial analysts, top-management) should determine some important parameters (e.g. potential losses, direct investment costs, internal and external domino accident probabilities) to provide the MPC with the required information such that it is able to calculate the different cost functions (see also Chap. 5).

This is a necessary step for the model, since only if all required confidential financial and operational information from all chemical plants in the multi-plant area is delivered to the Multi-Plant Council Data Administration, it is possible for the independent experts of the MPC Data Administration (cfr. Chap. 2 for issues related to the meaning of the MPC, its characteristics, its composition, its handling of confidential information, etc.) to draw aggregated conclusions. The expected damage for plant $i, i \in \varphi$, due to a domino accident is calculated on the basis of realized consequences of investment decisions by other plants (other than i) $j, j \neq i \in \varphi$, in the multi-plant area.

The MPC is able to calculate collaboration benefits for the plants by determining the difference between two scenarios. The first scenario is the To-Be situation (i.e. application of the CIA), whereas the second scenario represents the Business-As-Usual situation (i.e. without any cross-plant precaution investment). The MPC calculates the collaboration benefits and the maximum feasible cooperation that can be achieved among the companies by assuming that the individual plants deliver truthful information. The investment decision for the chemical plants in the Business-As-Usual situation depends on the criteria and perceptions, on which company prevention management bases its current prevention strategy choice. The Business-As-Usual situation can be either the non-cooperative case, or the case in which partial cooperation is present to some extent. One consequently needs to determine what are the criteria and perceptions on which prevention managers currently base their domino effects prevention investment decisions on. We refer to these criteria and perceptions as the 'decision factors' for company prevention management.

The human mind deals with uncertainty, risks, and decisions by resorting to a set of heuristics (Hubbart 2009). A heuristic is a sort of mental shortcut that in man's simpler, hunter-gatherer times probably sufficed for a variety of situations, and still does today. A related concept is *bias*, that is, a tendency to think and behave in a way that interferes with rationality and impartiality. We try to take into account the biases of prevention management and use them to obtain an idea of the mind-heuristic that is being used by decision makers to decide on cross-plant precautions. In summary, a wrong perception of reality leads to wrong, or at least suboptimal, decisions, and therefore the decision-makers' perception needs to be taken into account when elaborating on CIA.

Let us assume that decision-makers are risk-neutral and their risk attitude has a certainty equivalent equal to the expected value of an outcome. As before, n denotes the number of plants in the chemical cluster φ. Let q denote the number of plants ψ, which are cooperating on preventive measures. Without loss of generality, we assume that chemical plants have identical probabilities of internal and external domino accidents, i.e., $P_{\mathrm{ii}} = P_{\mathrm{jj}}$ and $P_{\mathrm{ji}} = P_{\mathrm{ki}}$ for all i and $j \neq k \neq i$. In this case, annual expected costs due to a domino accident initiated in a chemical plant $i \in \varphi\backslash\psi$ in case it decides to *Invest* in domino accident prevention or *Not Invest* are indicated by expressions (6.1–6.2), respectively,

$$u_i(\mathrm{I}) = c_i + (n - (q+1)) \cdot P_{\mathrm{ji}} L_i \cdot (1 - P_{\mathrm{ji}})^{n-(q+1)-1} \tag{6.1}$$

$$u_i(\mathrm{NI}) = P_{\mathrm{ii}} L_i (1 - P_{\mathrm{ji}})^{n-q-1} + (n - (q+1)) \cdot P_{\mathrm{ji}} L_i \cdot (1 - P_{\mathrm{ji}})^{n-(q+1)-1} \tag{6.2}$$

where $i \in \varphi\backslash\psi$ and $j \neq i \in \varphi\backslash\psi$.

Although these expressions hold for a cluster composed of chemical plants with identical probabilities of internal and external domino effects (that is, an 'homogeneous cluster'), the same decision factors are applicable to a multi-plant area in which probabilities of domino accidents differ across the plants (that is, a 'heterogeneous multi-plant area').

Observations over chemical plants in real industrial clusters show that prevention management perceives the probability of an external domino effect to be much lower than the probability of an internal domino effect. Since the latter (internal domino effect) probability is perceived as extremely low, there are—at very best—insufficient prevention investments for such accidents. Prevention management approximates external domino effect probability by zero ($P_{\mathrm{ji}} = 0$) in its perception. It follows from Eqs. (6.1–6.2) that the perception of domino risks are realized by individual plants in the following way (and it does not make a difference in this context whether the adjacent plant is identical or not).

$$u_i(\mathrm{I}) \approx c_i \tag{6.3}$$

$$u_i(\mathrm{NI}) \approx P_{ii}L_i. \tag{6.4}$$

These expressions demonstrate in a simplified way, which factors affect plants' decisions regarding domino prevention management. The reader should be aware that neither internal or external domino accident probabilities nor the common terms in Eqs. (6.1–6.2) are negligible in real industrial practice if one wants to take all types of risks into account (thus also type II and III risks) and if one wants to prevent true disasters, but that these approximations follow from individuals' perceptions. Expressions (6.3–6.4) show that the decision to invest or not in cross-company prevention from a prevention managers' point of view is dependent on whether or not it is cheaper to either invest in domino prevention measures (i.e., pay an investment c_i) or not (i.e., possibly encounter an expected damage resulting from an internal domino effect that equals $P_{\mathrm{ii}}L_i$). A chemical plant will therefore be inclined to invest in prevention measures if $c_i \leq P_{ii}L_i$, and will decide not to invest in case $c_i > P_{\mathrm{ii},}L_i$

This result supplements Proposition 4.1, where plants' interdependence through possible external domino effects is fully taken into account. It shows that if the investments are equal or larger than the expected direct loss from not investing in preventive measures, i.e., $c_i > P_{\mathrm{ii},}L_i$, then strategic interaction among chemical plants resembles the Prisoner's dilemma type of coordination game (Not Invest becomes a dominant strategy), where the only equilibrium is no cooperation. If investments are smaller or equal than expected direct loss from not investing in preventive measures, then the strategy profile (Invest, Invest) is a Nash equilibrium and cooperation between the companies is possible. However, even though (Invest, Invest) is a Nash equilibrium, the strategy *Invest* of either of the players is not necessarily dominant in the Stag hunt game, strategic interaction can lead to partial cooperation of chemical plants. These decision factors serve as important background knowledge for the CIA.

The expected costs of the chemical plants in the Business-As-Usual situation are compared by the MPC (using the CIA approach) with the expected costs in the To-Be situation. The expected costs in the To-Be situation are those in which the total expected losses of the cluster are minimized. This way, the collaboration benefits are maximized compared to the Business-As-Usual situation.

It should be noted that most probably in real industrial practice collaboration benefits are not at all maximized. A chemical plant obviously 'wins' in case expected costs in domino effect prevention in the To-Be situation are lower than its expected costs in the Business-As-Usual situation. A chemical plant facing lower expected costs in the Business-As-Usual situation compared to the To-Be situation, obviously faces a deficit. In a chemical industrial area, a plant should never lose as a consequence of cooperation. The MPC Data Administration needs to use the surpluses to compensate the deficits so that no plant individually loses compared to the Business-As-Usual situation. The remaining financial resources which are saved due to strategic collaboration among plants can then be divided among the plants belonging to the multi-plant area. It should be noted that the MPC may recommend how to use the collaboration benefits ('cooperative surplus' in game-theoretic terms) to the single plants. The individual expected costs are then again compared to the Business-As-Usual situation and the cooperation type (full-, partial- or no cooperation) of the cluster is determined. The CIA approach does not rely on external incentives such as subsidies or taxes or insurance fee incentives (possibly provided by authorities or by insurance companies respectively) to stimulate cooperation.[1] The suggested CIA can be used autonomously to stimulate cooperation among chemical plants in a cluster by means of redistributing possible financial benefits from collaboration. For this reason, it is not possible in every situation to stimulate cooperation by means of the proposed model. Some situations do require financial interference external to the cluster to induce collaboration between the plants. To illustrate the model's applicability, the next section gives examples of different cluster situations.

The starting point of the analysis conducted by the MPC is to investigate the strategic situation occurring when no supra-plant body would be present in the chemical industrial area. As indicated before, we assume here that company *i*'s prevention management bases its decision whether or not to invest in external domino prevention measures basically on the difference between the investments c_i and the expected losses caused by possible internal domino effect $L_i P_{ii}$.

We therefore distinguish four situations in the Cooperative Incentives Approach:

1. It is for every plant in the cluster less costly to invest in prevention measures than to risk an internal domino effect;
2. It is for some plants (but not for all of them) less costly to invest in prevention measures than to risk an internal domino effect. However, the collaboration benefits of the chemical cluster are maximized in a strategic situation where all plants invest in domino effect prevention (we call this "full cooperation");
3. It is for some plants (but not for all of them) less costly to invest in domino prevention measures than to risk the costs and probabilities of an internal

[1] The MPC obviously needs to have an internal working budget for its base cost of existence (man hours, office, etc.), but we assume that there is no external 'incentive budget' available that can be used by the MPC to stimulate domino prevention collaboration between companies.

domino effect. However, a situation where all plants fully cooperate with regard to domino effect prevention does not maximize the collaboration benefits of the chemical cluster (this can lead to a situation of full cooperation or of partial cooperation);

4. It is for every plant in the cluster more expensive to invest in prevention measures than to bare with possible risks related to an internal domino effect.

The implications and the resulting type of cooperation of these situations are illustrated in the next section.

6.3 Applying the CIA Approach: An Illustrative Example

We assume a chemical industrial area consisting of five heterogeneous companies (*Plant A, Plant B, Plant C, Plant D, Plant E*), Table 6.1 provides an overview of the various theoretical cost functions of *Plant A* in all possible strategic situations.[2] We also refer to Reniers et al. (2012).

We further show the simplicity, user-friendliness, and usefulness as well as the limitations of the CIA approach by means of four examples. In these examples, the five chemical plants each have specific parameters. In a cluster composed of 5 plants, we distinguish 32 (i.e., 2^5) strategic situations for every plant and therefore a total of 164 expected costs (i.e. 5×2^5). To improve comparability over the three situations in Table 6.2 we use identical internal and external probabilities as well as identical potential losses for each of the four 'Illustrative examples/Situations', described and discussed hereafter.

In Table 6.2, the plants designated in the rows initiate the domino effects, while the plants designated in the columns suffer losses from the initiating plant. For example, the probability that a domino accident in Plant C causes damage to Plant E is: $P_{CE} = 1.20 \times 10^{-6}$ per year.

6.3.1 *Illustrative Example/Situation (1)*

Situation (1) describes an example in which the *single company invest*—strategy results in the lowest possible investment cost for all plants. We assume thus that every plant directly invests in domino prevention measures and does not risk a domino effect, because it is in the perception of company prevention management less expensive to invest than to not invest. Assume the following investment costs for five chemical plants belonging to the same industrial area:

[2] A strategic situation is defined as a simultaneous strategy selection, one for each plant. Each strategic situation results in a different pay-off or investment cost for each player.

Table 6.1 Expected costs of *Plant A* in a 5-plant chemical multi-plant area

Strategic situation (*A,B,C,D,E*)	Expected costs of *Plant A*
(I, I, I, I, I)	c_A
(I, I, I, I, NI)	$c_A + P_{\text{EA}}L_A$
(I, I, I, NI, I)	$c_A + P_{\text{DA}}L_A$
(I, I, NI, I, I)	$c_A + P_{\text{CA}}L_A$
(I, NI, I, I, I)	$c_A + P_{\text{BA}}L_A$
(I, I, I, NI, NI)	$c_A + P_{\text{DA}}L_A(1 - P_{\text{EA}}) + P_{\text{EA}}L_A(1 - P_{\text{DA}})$
(I, I, NI, I, NI)	$c_A + P_{\text{CA}}L_A(1 - P_{\text{EA}}) + P_{\text{EA}}L_A(1 - P_{\text{CA}})$
(I, NI, I, I, NI)	$c_A + P_{\text{BA}}L_A(1 - P_{\text{EA}}) + P_{\text{EA}}L_A(1 - P_{\text{BA}})$
(I, I, NI, NI, I)	$c_A + P_{\text{CA}}L_A(1 - P_{\text{DA}}) + P_{\text{DA}}L_A(1 - P_{\text{CA}})$
(I, NI, I, NI, I)	$c_A + P_{\text{BA}}L_A(1 - P_{\text{DA}}) + P_{\text{DA}}L_A(1 - P_{\text{BA}})$
(I, NI, NI, I, I)	$c_A + P_{\text{BA}}L_A(1 - P_{\text{CA}}) + P_{\text{CA}}L_A(1 - P_{\text{BA}})$
(I, I, NI, NI, NI)	$c_A + P_{\text{CA}}L_A(1 - P_{\text{DA}})(1 - P_{\text{EA}}) + P_{\text{DA}}L_A(1 - P_{\text{EA}})(1 - P_{\text{CA}}) + P_{\text{EA}}L_A(1 - P_{\text{CA}})\ (1 - P_{D,A})$
(I, NI, I, NI, NI)	$c_A + P_{\text{BA}}L_A(1 - P_{\text{DA}})(1 - P_{\text{EA}}) + P_{\text{DA}}L_A(1 - P_{\text{EA}})(1 - P_{\text{BA}}) + P_{\text{EA}}L_A(1 - P_{\text{BA}})\ (1 - P_{\text{DA}})$
(I, NI, NI, I, NI)	$c_A + P_{\text{BA}}L_A(1 - P_{\text{CA}})(1 - P_{\text{EA}}) + P_{\text{CA}}L_A(1 - P_{\text{EA}})(1 - P_{\text{BA}}) + P_{\text{EA}}L_A(1 - P_{\text{BA}})\ (1 - P_{\text{CA}})$
(I, NI, NI, NI, I)	$c_A + P_{\text{BA}}L_A(1 - P_{\text{CA}})(1 - P_{\text{DA}}) + P_{\text{CA}}L_A(1 - P_{\text{DA}})(1 - P_{\text{BA}}) + P_{\text{DA}}L_A(1 - P_{\text{BA}})(1 - P_{C,A})$
(I, NI, NI, NI, NI)	$c_A + P_{\text{BA}}L_A(1 - P_{\text{CA}})(1 - P_{\text{DA}})(1 - P_{\text{EA}}) + P_{\text{CA}}L_A(1 - P_{\text{BA}})(1 - P_{\text{DA}})(1 - P_{\text{EA}}) + P_{\text{DA}}L_A(1 - P_{\text{BA}})(1 - P_{\text{CA}})$ $(1 - P_{\text{EA}}) + P_{\text{EA}}L_A(1 - P_{\text{BA}})(1 - P_{\text{CA}})(1 - P_{\text{DA}})$
(NI, NI, NI, NI, NI)	$P_{\text{AA}}L_A(1 - P_{\text{BA}})(1 - P_{\text{CA}})(1 - P_{\text{DA}})(1 - P_{\text{EA}}) + P_{\text{BA}}L_A(1 - P_{\text{CA}})(1 - P_{\text{DA}})(1 - P_{\text{EA}})(1 - P_{A,A}) + P_{\text{CA}}L_A(1 - P_{\text{BA}})$ $(1 - P_{\text{DA}})(1 - P_{\text{EA}})(1 - P_{\text{AA}}) + P_{\text{DA}}L_A(1 - P_{\text{BA}})(1 - P_{\text{CA}})(1 - P_{\text{EA}})(1 - P_{\text{AA}}) + P_{\text{EA}}L_A(1 - P_{\text{BA}})(1 - P_{\text{CA}})(1 - P_{\text{DA}})$ $(1 - P_{\text{AA}})$
(NI, NI, NI, NI, I)	$P_{\text{AA}}L_A(1 - P_{\text{BA}})(1 - P_{\text{CA}})(1 - P_{\text{DA}}) + P_{\text{BA}}L_A(1 - P_{\text{CA}})(1 - P_{\text{DA}})(1 - P_{\text{AA}}) + P_{\text{CA}}L_A(1 - P_{\text{BA}})(1 - P_{\text{DA}})$ $(1 - P_{\text{AA}}) + P_{\text{DA}}L_A(1 - P_{\text{BA}})(1 - P_{\text{CA}})(1 - P_{\text{AA}})$
(NI, NI, NI, I, NI)	$P_{\text{AA}}L_A(1 - P_{\text{BA}})(1 - P_{\text{CA}})(1 - P_{\text{EA}}) + P_{\text{BA}}L_A(1 - P_{\text{CA}})(1 - P_{\text{EA}})(1 - P_{\text{AA}}) + P_{\text{CA}}L_A(1 - P_{\text{BA}})(1 - P_{\text{EA}})$ $(1 - P_{\text{AA}}) + P_{\text{EA}}L_A(1 - P_{\text{BA}})(1 - P_{\text{CA}})(1 - P_{\text{AA}})$

(continued)

Table 6.1 (continued)

Strategic situation (*A,B,C,D,E*)	Expected costs of *Plant A*
(NI, NI, I, NI, NI)	$P_{AA}L_A(1-P_{BA})(1-P_{DA})(1-P_{EA})+P_{BA}L_A(1-P_{DA})(1-P_{EA})(1-P_{AA})+P_{DA}L_A(1-P_{BA})$ $(1-P_{EA})(1-P_{AA})+P_{EA}L_A(1-P_{BA})(1-P_{DA})(1-P_{AA})$
(NI, I, NI, NI, NI)	$P_{AA}L_A(1-P_{CA})(1-P_{DA})(1-P_{EA})+P_{CA}L_A(1-P_{DA})(1-P_{EA})(1-P_{AA})+P_{DA}L_A(1-P_{CA})$ $(1-P_{EA})(1-P_{AA})+P_{EA}L_A(1-P_{CA})(1-P_{DA})(1-P_{AA})$
(NI, NI, NI, I, I)	$P_{AA}L_A(1-P_{BA})(1-P_{CA})+P_{BA}L_A(1-P_{CA})(1-P_{AA})+P_{CA}L_A(1-P_{BA})(1-P_{AA})$
(NI, NI, I, NI, I)	$P_{AA}L_A(1-P_{BA})(1-P_{DA})+P_{BA}L_A(1-P_{DA})(1-P_{AA})+P_{DA}L_A(1-P_{BA})(1-P_{AA})$
(NI, I, NI, NI, I)	$P_{AA}L_A(1-P_{CA})(1-P_{DA})+P_{CA}L_A(1-P_{DA})(1-P_{AA})+P_{DA}L_A(1-P_{CA})(1-P_{AA})$
(NI, NI, I, I, NI)	$P_{AA}L_A(1-P_{BA})(1-P_{EA})+P_{BA}L_A(1-P_{EA})(1-P_{AA})+P_{EA}L_A(1-P_{BA})(1-P_{AA})$
(NI, I, NI, I, NI)	$P_{AA}L_A(1-P_{CA})(1-P_{EA})+P_{CA}L_A(1-P_{EA})(1-P_{AA})+P_{EA}L_A(1-P_{CA})(1-P_{AA})$
(NI, I, I, NI, NI)	$P_{AA}L_A(1-P_{DA})(1-P_{EA})+P_{DA}L_A(1-P_{EA})(1-P_{AA})+P_{EA}L_A(1-P_{DA})(1-P_{AA})$
(NI, NI, I, I, I)	$P_{AA}L_A(1-P_{BA})+P_{BA}L_A(1-P_{AA})$
(NI, I, NI, I, I)	$P_{AA}L_A(1-P_{CA})+P_{CA}L_A(1-P_{AA})$
(NI, I, I, NI, I)	$P_{AA}L_A(1-P_{DA})+P_{DA}L_A(1-P_{AA})$
(NI, I, I, I, NI)	$P_{AA}L_A(1-P_{EA})+P_{EA}L_A(1-P_{AA})$
(NI, I, I, I, I)	$P_{AA}L_A$

Table 6.2 Losses and Domino Accident Probabilities (DAP) for the illustrative examples

DAP (times/year)	*Plant A*	*Plant B*	*Plant C*	*Plant D*	*Plant E*	*Potential losses (€)*
Plant A	1.10×10^{-4}	2.50×10^{-6}	3.40×10^{-6}	4.80×10^{-6}	0.90×10^{-6}	1.60×10^{8}
Plant B	2.30×10^{-6}	2.10×10^{-4}	4.30×10^{-6}	7.60×10^{-6}	8.80×10^{-6}	1.10×10^{8}
Plant C	6.50×10^{-6}	5.40×10^{-6}	3.80×10^{-4}	3.20×10^{-6}	1.20×10^{-6}	1.35×10^{8}
Plant D	6.60×10^{-6}	3.80×10^{-6}	7.60×10^{-6}	3.60×10^{-4}	2.20×10^{-6}	2.20×10^{8}
Plant E	8.10×10^{-6}	1.60×10^{-6}	2.50×10^{-6}	8.90×10^{-6}	7.50×10^{-4}	1.90×10^{8}

Let investment in preventive measures for each of the plants be as follows:

Plant A: $c_A =$ 16,000 €/year,
Plant B: $c_B =$ 22,000 €/year,
Plant C: $c_C =$ 48,000 €/year,
Plant D: $c_D =$ 68,000 €/year,
Plant E: $c_E =$ 103,000 €/year.

In this situation, the corresponding Business-As-Usual situation for these parameters is equal to the strategic situation (I, I, I, I, I). The MPC Data Administration carries out the CIA simulation and calculates the expected costs for every plant in every strategic situation according to game-theoretic model presented in Chaps. 4 and 5. This result is shown in Table 6.3.

Table 6.3 indicates that the To-Be situation in this example is actually the situation in which all plants do invest in external domino effects precaution, that is, the Business-As-Usual situation. Since the Business-As-Usual situation and the To-Be situation are identical in this situation, there is obviously no real use for the CIA approach to redistribute costs to induce collaboration. In this example, the prevention managers would individually choose to invest and by doing so already inadvertently maximize the collaboration benefits.

6.3.2 Illustrative Example/Situation (2)

In this situation, it is cheaper for some plants, but not for all, to invest in prevention measures rather than to risk the possible losses related to an internal domino effect. The difference with the previous example is that full cooperation is not Nash equilibrium of this game. Even though the all-invest outcome results in the lowest expected costs (Pareto efficient outcome), it is not stable due to incentives of some plants to deviate. To make Pareto efficient outcome stable, the *rules of the game* need to be changed.

The Multi-Plant Council can gather and assess information concerning plants' potential damages and probabilities of direct and indirect accidents. Resulting analysis shows that cooperation among the plants is possible without use of financial

Table 6.3 Expected costs for situation (1)

Strategic situations	*Plant A*	*Plant B*	*Plant C*	*Plant D*	*Plant E*	Total expected costs
(I, I, I, I, I)	**16,000.00**	**22,000.00**	**48,000.00**	**68,000.00**	**103,000.00**	**257,000.00**
(I, I, I, I, NI)	16,144.00	22,968.00	48,162.00	68,484.00	142,500.00	298,258.00
(I, I, I, NI, I)	16,768.00	22,836.00	48,432.00	79,200.00	104,691.00	271,927.00
(I, I, NI, I, I)	16,544.00	22,473.00	51,300.00	69,672.00	103,475.00	263,464.00
(I, NI, I, I, I)	16,400.00	23,100.00	48,729.00	68,836.00	103,304.00	260,369.00
(I, I, I, NI, NI)	16,912.00	23,803.99	48,594.00	79,683.65	144,188.46	313,182.10
(I, I, NI, I, NI)	16,688.00	23,440.99	51,461.88	70,155.99	142,974.29	304,721.15
(I, NI, I, I, NI)	16,544.00	24,067.59	48,891.00	69,320.00	142,803.54	301,626.13
(I, I, NI, NI, I)	17,311.99	23,308.99	51,731.67	80,870.80	105,165.99	278,389.45
(I, NI, I, NI, I)	17,168.00	23,935.65	49,161.00	80,035.40	104,994.99	275,295.03
(I, NI, NI, I, I)	16,944.00	23,572.80	52,028.45	70,507.99	103,779.00	266,832.23
(I, I, NI, NI, NI)	17,455.99	24,276.97	51,893.55	81,354.44	144,662.74	319,643.69
(I, NI, I, NI, NI)	17,311.99	24,903.23	49,322.99	80,519.05	144,492.00	316,549.26
(I, NI, NI, I, NI)	17,088.00	24,540.39	52,190.32	70,991.98	143,277.83	308,088.51
(I, NI, NI, NI, I)	17,711.99	24,408.44	52,460.11	81,706.18	105,469.98	281,756.71
(I, NI, NI, NI, NI)	17,855.99	25,376.01	52,621.99	82,189.82	144,966.28	323,010.09
(NI, NI, NI, NI, NI)	**19,455.58**	**25,628.90**	**53,498.80**	**83,640.74**	**146,502.93**	**328,726.94**
(NI, NI, NI, NI, I)	19,311.61	24,661.33	53,336.93	83,157.10	107,008.94	287,475.92
(NI, NI, NI, I, NI)	18,687.76	24,793.27	53,067.14	72,443.94	144,814.51	313,806.62
(NI, NI, I, NI, NI)	18,911.71	25,156.11	50,200.48	81,969.98	146,028.66	322,266.94
(NI, I, NI, NI, NI)	19,055.67	24,529.96	52,770.37	82,805.37	146,199.40	325,360.77
(NI, NI, NI, I, I)	18,543.79	23,825.69	52,905.27	71,959.95	105,317.99	272,552.69
(NI, NI, I, NI, I)	18,767.74	24,188.54	50,038.48	81,486.34	106,533.96	281,015.06
(NI, I, NI, NI, I)	18,911.71	23,561.99	52,608.50	82,321.73	106,704.96	284,108.88
(NI, NI, I, I, NI)	18,143.88	24,320.48	49,768.49	70,771.98	144,340.23	307,345.06
(NI, I, NI, I, NI)	18,287.85	23,693.99	52,338.71	71,607.96	144,510.97	310,439.48
(NI, I, I, NI, NI)	18,511.80	24,056.98	49,471.49	81,134.60	145,725.13	318,899.99
(NI, NI, I, I, I)	17,999.91	23,352.89	49,606.49	70,287.99	104,843.00	266,090.28
(NI, I, NI, I, I)	18,143.88	22,726.00	52,176.83	71,123.98	105,013.99	269,184.68
(NI, I, I, NI, I)	18,367.83	23,089.00	49,309.49	80,650.95	106,229.97	277,647.25
(NI, I, I, I, NI)	17,743.97	23,221.00	49,039.50	69,935.99	144,036.69	303,977.15
(NI, I, I, I, I)	17,600.00	22,253.00	48,877.50	69,452.00	104,539.00	262,721.50

incentives from the MPC. However, *information incentives* are necessary to change the rules of the game in order to establish the desirable (I, I, I, I, I)-situation.

Let investment in preventive measures for each of the plants be as follows:

Plant A: $c_A = 21{,}000$ €/year,
Plant B: $c_B = 22{,}000$ €/year,
Plant C: $c_C = 53{,}000$ €/year,
Plant D: $c_D = 68{,}000$ €/year,
Plant E: $c_E = 103{,}000$ €/year.

Table 6.4 Expected costs for situation (2)

Strategic situation	*Plant A*	*Plant B*	*Plant C*	*Plant D*	*Plant E*	Total expected costs
(I, I, I, I, I)	**21,000.00**	**22,000.00**	**53,000.00**	**68,000.00**	**103,000.00**	**267,000.00**
(I, I, I, I, NI)	21,144.00	22,968.00	53,162.00	68,484.00	142,500.00	308,258.00
(I, I, I, NI, I)	21,768.00	22,836.00	53,432.00	79,200.00	104,691.00	281,927.00
(I, I, NI, I, I)	21,544.00	22,473.00	51,300.00	69,672.00	103,475.00	268,464.00
(I, NI, I, I, I)	21,400.00	23,100.00	53,729.00	68,836.00	103,304.00	270,369.00
(I, I, I, NI, NI)	21,912.00	23,803.99	53,594.00	79,683.65	144,188.46	323,182.10
(I, I, NI, I, NI)	21,688.00	23,440.99	51,461.88	70,155.99	142,974.29	309,721.15
(I, NI, I, I, NI)	21,544.00	24,067.59	53,891.00	69,320.00	142,803.54	311,626.13
(I, I, NI, NI, I)	22,311.99	23,308.99	51,731.67	80,870.80	105,165.99	283,389.45
(I, NI, I, NI, I)	22,168.00	23,935.65	54,161.00	80,035.40	104,994.99	285,295.03
(I, NI, NI, I, I)	21,944.00	23,572.80	52,028.45	70,507.99	103,779.00	271,832.23
(I, I, NI, NI, NI)	22,455.99	24,276.97	51,893.55	81,354.44	144,662.74	324,643.69
(I, NI, I, NI, NI)	22,311.99	24,903.23	54,322.99	80,519.05	144,492.00	326,549.26
(I, NI, NI, I, NI)	22,088.00	24,540.39	52,190.32	70,991.98	143,277.83	313,088.51
(I, NI, NI, NI, I)	22,711.99	24,408.44	52,460.11	81,706.18	10,5469.98	286,756.71
(I, NI, NI, NI, NI)	22,855.99	25,376.01	52,621.99	82,189.82	144,966.28	328,010.09
(NI, NI, NI, NI, NI)	**19,455.58**	**25,628.90**	**53,498.80**	**83,640.74**	**146,502.93**	**328,726.94**
(NI, NI, NI, NI, I)	19,311.61	24,661.33	53,336.93	83,157.10	107,008.94	287,475.92
(NI, NI, NI, I, NI)	18,687.76	24,793.27	53,067.14	72,443.94	144,814.51	313,806.62
(NI, NI, I, NI, NI)	18,911.71	25,156.11	55,200.48	81,969.98	146,028.66	327,266.94
(NI, I, NI, NI, NI)	19,055.67	24,529.96	52,770.37	82,805.37	146,199.40	325,360.77
(NI, NI, NI, I, I)	18,543.79	23,825.69	52,905.27	71,959.95	105,317.99	272,552.69
(NI, NI, I, NI, I)	18,767.74	24,188.54	55,038.48	81,486.34	106,533.96	286,015.06
(NI, I, NI, NI, I)	18,911.71	23,561.99	52,608.50	82,321.73	106,704.96	284,108.88
(NI, NI, I, I, NI)	18,143.88	24,320.48	54,768.49	70,771.98	144,340.23	312,345.06
(NI, I, NI, I, NI)	18,287.85	23,693.99	52,338.71	71,607.96	144,510.97	310,439.48
(NI, I, I, NI, NI)	18,511.80	24,056.98	54,471.49	81,134.60	145,725.13	323,899.99
(NI, NI, I, I, I)	17,999.91	23,352.89	54,606.49	70,287.99	104,843.00	271,090.28
(NI, I, NI, I, I)	18,143.88	22,726.00	52,176.83	71,123.98	105,013.99	269,184.68
(NI, I, I, NI, I)	18,367.83	23,089.00	54,309.49	80,650.95	106,229.97	282,647.25
(NI, I, I, I, NI)	17,743.97	23,221.00	54,039.50	69,935.99	144,036.69	308,977.15
(NI, I, I, I, I)	17,600.00	22,253.00	53,877.50	69,452.00	104,539.00	267,721.50

Table 6.4 shows the total annual expected costs for every plant in all possible situations with the cross-plant information available to the MPC Data Administration. These numbers show that Plant A and Plant C will choose not to invest in prevention measures.

Plant A: $c_A = 21{,}000$ €/year $> P_{AA}L_A = 17{,}600$ €/year
Plant C: $c_C = 53{,}000$ €/year $> P_{CC}L_C = 51{,}300$ €/year

The Business-As-Usual situation that would result without the use of the CIA approach is therefore the strategic situation (NI, I, NI, I, I). Table 6.4 shows that collaboration benefits are maximized in the all-invest situation (i.e., the To-Be

Table 6.5 Surplus (+) and deficit (−) per plant in situation (2)

Strategic situation	*Plant A*	*Plant B*	*Plant C*	*Plant D*	*Plant E*	Total costs
(NI, I, NI, I, I)	18,143.88	22,726.00	52,176.83	71,123.98	105,013.99	269,184.68
(I, I, I, I, I)	21,000.00	22,000.00	53,000.00	68,000.00	103,000.00	267,000.00
Surplus/deficit:	−2,856.12	+726.00	−823.17	+3,123.98	+2,013.99	+2,184.68

situation). It should be noted that for the calculation of collaboration benefits, expected losses from internal as well as external domino effects are taken into account.

After calculating all investment costs, an analysis of the difference between the Business-As-Usual situation and the To-Be situation is performed, as shown in Table 6.5.

The deficits of Plant A and Plant C will be compensated by the surpluses of Plant B, Plant D, and Plant E in a linear manner by the MPC.

Total Surplus (Plant B + Plant D + Plant E) = € 5,863.97
Total Deficit (Plant A + Plant C) = € 3,679.29
→ **Difference** (Surplus−Deficit) = € 2,184.68

- *Plant B* pays 12.38 % (i.e. 726/5,863.97) of the deficit of *Plant A* and *Plant C*:

Compensation *Plant B* € 353.60 (i.e., 12.38 % of € 2856.12) to *Plant A*
€ 101.91 (i.e., 12.38 % of € 823.17) to *Plant C*

- *Plant D* pays 53.27 % (i.e., 3,123.98/5,863.97) of the deficit of *Plant A* and *Plant C*:

Compensation *Plant D* € 1521.46 (i.e., 53.27 % of € 2856.12) to *Plant A*
€ 438.50 (i.e. 53.27 % of € 823.17) to *Plant C*

- *Plant E* pays 34.35 % (i.e., 2,013.99/5,863.97) of the deficit of *Plant A* and *Plant C*:

Compensation *Plant E* € 981.08 (i.e., 34.35 % of € 2856.12) to *Plant A*
€ 282.76 (i.e., 34.35 % of € 823.17) to *Plant C.*

Next, the remaining surplus of € 2,184.68 as a consequence of the maximization of collaboration benefits can be redistributed according to a mechanism agreed upon within the industrial cluster and the MPC. For the sake of simplicity and for illustrating our example, we redistribute it linearly in this book. This implies that each plant's expected costs is reduced by € 436.94 (i.e., € 2,184.68/5). The annual expected costs of all plants in the To-Be situation can then be found in Table 6.6.

Table 6.6 Expected costs in situation (2)

Strategic situation	*Plant A*	*Plant B*	*Plant C*	*Plant D*	*Plant E*	Total expected costs
(I, I, I, I, I)*	17,706.94	22,289.06	51,739.89	70,687.04	104,577.05	267,000.00

It is important to note that the use of the CIA in this situation minimizes the total expected losses for the cluster and therefore maximizes the collaboration benefits among the chemical plants. By redistributing these benefits, every chemical plant in the multi-plant area faces lower (or at least equal) annual expected costs in the To-Be-situation in comparison to the costs in the Business-As-Usual situation.

In this example, the expected costs for plants A and C in the To-Be situation (17,706.94 €/year and 51,739.89 €/year, respectively) is still higher than the investments (17,600 €/year and 51,300 €/year, respectively), perceived as maximum. However, as already mentioned, the *perceived costs of investments* are merely used to identify in which strategic collaboration situation the cluster will find itself, based on perceptions by prevention management. In this case, the situation would be (NI, I, NI, I, I). In that situation, the expected costs for plants A and C are calculated to be 18,143.88 €/year and 52,176.83 €/year, respectively, and thus higher than the expected costs in the To-Be situation.

6.3.3 *Illustrative Example/Situation (3)*

In this case, it is cheaper for some but not for all plants to invest in prevention measures rather than risk initiating a domino effect. In this illustrative simulation, the direct investments cost of Plant B are significantly increased, resulting in a situation in which the all-invest strategy no longer minimizes the total expected costs for the cluster. Let us assume the following investments:

Plant A: $C_A = 21{,}000$ €/year,
Plant B: $C_B = 30{,}000$ €/year,
Plant C: $C_C = 53{,}000$ €/year,
Plant D: $C_D = 68{,}000$ €/year,
Plant E: $C_E = 103{,}000$ €/year.

The total annual expected costs for all plants as well as for the studied industrial area are given in Table 6.7. This figure indicates that the strategic situation (I, NI, I, I, I) minimizes the total expected costs for the cluster.

From Table 6.7, we can deduce that the Business-As-Usual situation would be the strategic situation (NI, NI, NI, I, I) since it is cheaper for Plant A, Plant B, and Plant C to risk initiating a domino effect rather than to invest in prevention measures:

Plant A: $C_A = € \ 21{,}000 > P_{A,A}L_A = € \ 17{,}600$
Plant B: $C_B = € \ 30{,}000 > P_{B,B}L_B = € \ 23{,}100$
Plant C: $C_C = € \ 53{,}000 > P_{C,C}L_C = € \ 51{,}300$

Table 6.7 Expected costs for situation (3)

Strategic situations	*Plant A*	*Plant B*	*Plant C*	*Plant D*	*Plant E*	Total expected costs
(I, I, I, I, I)	21,000.00	30,000.00	53,000.00	68,000.00	103,000.00	275,000.00
(I, I, I, I, NI)	21,144.00	30,968.00	53,162.00	68,484.00	142,500.00	316,258.00
(I, I, I, NI, I)	21,768.00	30,836.00	53,432.00	79,200.00	104,691.00	289,927.00
(I, I, NI, I, I)	21,544.00	30,473.00	51,300.00	69,672.00	103,475.00	276,464.00
(I, NI, I, I, I)	21,400.00	23,100.00	53,729.00	68,836.00	103,304.00	**270,369.00**
(I, I, I, NI, NI)	21,912.00	31,803.99	53,594.00	79,683.65	144,188.46	331,182.10
(I, I, NI, I, NI)	21,688.00	31,440.99	51,461.88	70,155.99	142,974.29	317,721.15
(I, NI, I, I, NI)	21,544.00	24,067.59	53,891.00	69,320.00	142,803.54	311,626.13
(I, I, NI, NI, I)	22,311.99	31,308.99	51,731.67	80,870.80	105,165.99	291,389.45
(I, NI, I, NI, I)	22,168.00	23,935.65	54,161.00	80,035.40	104,994.99	285,295.03
(I, NI, NI, I, I)	21,944.00	23,572.80	52,028.45	70,507.99	103,779.00	271,832.23
(I, I, NI, NI, NI)	22,455.99	32,276.97	51,893.55	81,354.44	144,662.74	332,643.69
(I, NI, I, NI, NI)	22,311.99	24,903.23	54,322.99	80,519.05	144,492.00	326,549.26
(I, NI, NI, I, NI)	22,088.00	24,540.39	52,190.32	70,991.98	143,277.83	313,088.51
(I, NI, NI, NI, I)	22,711.99	24,408.44	52,460.11	81,706.18	105,469.98	286,756.71
(I, NI, NI, NI, NI)	22,855.99	25,376.01	52,621.99	82,189.82	144,966.28	328,010.09
(NI, NI, NI, NI, NI)	19,455.58	25,628.90	53,498.80	83,640.74	146,502.93	328,726.94
(NI, NI, NI, NI, I)	19,311.61	24,661.33	53,336.93	83,157.10	107,008.94	287,475.92
(NI, NI, NI, I, NI)	18,687.76	24,793.27	53,067.14	72,443.94	144,814.51	313,806.62
(NI, NI, I, NI, NI)	18,911.71	25,156.11	55,200.48	81,969.98	146,028.66	327,266.94
(NI, I, NI, NI, NI)	19,055.67	32,529.96	52,770.37	82,805.37	146,199.40	333,360.77
(NI, NI, NI, I, I)	18,543.79	23,825.69	52,905.27	71,959.95	105,317.99	272,552.69
(NI, NI, I, NI, I)	18,767.74	24,188.54	55,038.48	81,486.34	106,533.96	286,015.06
(NI, I, NI, NI, I)	18,911.71	31,561.99	52,608.50	82,321.73	106,704.96	292,108.88
(NI, NI, I, I, NI)	18,143.88	24,320.48	54,768.49	70,771.98	144,340.23	312,345.06
(NI, I, NI, I, NI)	18,287.85	31,693.99	52,338.71	71,607.96	144,510.97	318,439.48
(NI, I, I, NI, NI)	18,511.80	32,056.98	54,471.49	81,134.60	145,725.13	331,899.99
(NI, NI, I, I, I)	17,999.91	23,352.89	54,606.49	70,287.99	104,843.00	271,090.28
(NI, I, NI, I, I)	18,143.88	30,726.00	52,176.83	71,123.98	105,013.99	277,184.68
(NI, I, I, NI, I)	18,367.83	31,089.00	54,309.49	80,650.95	106,229.97	290,647.25
(NI, I, I, I, NI)	17,743.97	31,221.00	54,039.50	69,935.99	144,036.69	316,977.15
(NI, I, I, I, I)	17,600.00	30,253.00	53,877.50	69,452.00	104,539.00	275,721.50

The To-Be situation in this example, as Table 6.7 shows, would be the strategic situation (I, NI, I, I, I). Table 6.8 shows the surpluses and deficits for all plants as a consequence of maximizing collaboration benefits.

The Multi-Plant Council will first use the surpluses of Plant B, Plant D, and Plant E to compensate the deficits of Plant A and Plant C. Second, the additional collaboration benefits (i.e., € 2,183.69) are redistributed linearly among the chemical plants. The resulting annual expected costs per plant are given in Table 6.9.

Table 6.8 Surplus (+) and deficit (−) per plant in situation (3)

Strategic situation	*Plant A*	*Plant B*	*Plant C*	*Plant D*	*Plant E*	Total expected costs
(NI, NI, NI, I, I)	18,543.79	23,825.69	52,905.27	71,959.95	105,317.99	272,552.69
(I, NI, I, I, I)	21,400.00	23,100.00	53,729.00	68,836.00	103,304.00	270,369.00
Surplus/deficit	−2,856.21	+725.69	−823.73	+3,123.95	+2,013.99	+2,183.69

Table 6.9 Investment costs per plant in strategic situation (I, NI, I, I, I)*

Strategic situation	*Plant A*	*Plant B*	*Plant C*	*Plant D*	*Plant E*	Total expected costs
(I, NI, I, I, I)*	18,107.05	23,388.95	52,468.53	71,523.21	104,881.25	270,369.00

Table 6.10 Expected costs per plant in the To-Be situation

Strategic situation	*Plant A*	*Plant B*	*Plant C*	*Plant D*	*Plant E*	Total expected costs
(I, NI, I, I, I)**	17,999.91	23,571.36	52,028.45	71,705.62	105,063.66	270,369.00

From Table 6.7 we can deduce that both Plant A and Plant C will not be persuaded to invest in prevention measures when facing the expected costs given in Table 6.9 since:

<u>*Plant A*</u>: u_A (NI, NI, I, I, I) = € 17,999.91 < u_A (I, NI, I, I, I)* = € 18,107.05
<u>*Plant C*</u>: u_C (I, NI, NI, I, I) = € 52,028.45 < u_C (I, NI, I, I, I)* = € 52,468.53

Therefore, an additional incentive of € 107.14 (= € 18,107.05 − € 17,999.91) and an additional incentive of € 440.08 (= € 52,468.53 − € 52,028.45) is necessary to shift Plant A, respectively, Plant C from the 'not-invest'- to the 'invest'-strategy. We propose that a Multi-Plant Council can obtain these incentives by means of collecting them from the other chemical plants in the cluster. We assume that these plants are willing to do this as long as they still benefit in comparison to the Business-As-Usual situation. Table 6.10 shows the total annual expected costs per plant as a consequence of fairly distributing collaboration benefits and additional incentives for Plant A and Plant C.

From Tables 6.7 and 6.10, we can now deduce that every plant benefits as a consequence of the CIA:

<u>*Plant A*</u>: u_A (I, NI, I, I, I)** = € 17,999.91 ≤ u_A (NI, NI, I, I, I) = € 17,999.91
<u>*Plant B*</u>: u_B (I, NI, I, I, I)** = € 23,571.36 ≤ u_B (I, I, I, I, I) = € 30,000
<u>*Plant C*</u>: u_C (I, NI, I, I, I)** = € 52,028.45 ≤ u_C (I, NI, NI, I, I) = € 52,028.45
<u>*Plant D*</u>: u_D (I, NI, I, I, I)** = € 71,705.62 ≤ u_D (I, NI, I, NI, I) = € 80,035.40
<u>*Plant E*</u>: u_E (I, NI, I, I, I)** = € 105,063.66 ≤ u_E (I, NI, I, I, NI) = € 142,803.54

According to the CIA approach, the all-invest situation is not a feasible (rational) outcome in this situation. This example, however, demonstrates that a situation of partial cooperation can be induced, such that it becomes the end outcome. The all-invest situation would only be possible by means of external incentives (that is, taxes or subsidies imposed or granted by authorities or insurance companies).

Let us now discuss a theoretical exercise (not being part of the CIA approach) and determine the possibility to achieve an all-invest situation after all for situation (3). In order to tip Plant B from the 'not-invest'- to the 'invest'-strategy, an external incentive (subsidy or tax) of at least € 6,428.64 is needed.

Subsidy (S) or insurance fee reduction for plant B:
$S_B \geq (C_B - P_{B,B}L_B)$
$S_B \geq$ € 6,428.64 (= (€ 30,000 − € 23,571.36))
Tax (T) or insurance fee increase for plant B:
$C_B \leq (T_B + P_{B,B}L_B)$
€ 30,000 $\leq$ (T_B + € 23,571.36)

6.3.4 Illustrative Example/Situation (4)

In this example, it is individually cheaper for all plants not to invest in domino effect prevention measures. We assume the following expected costs and compare them to the implicit impact costs.

Plant A: C_A = 20,000 €/year > $P_{AA}L_A$ = 17,600 €/year
Plant B: C_B = 25,000 €/year > $P_{BB}L_B$ = 23,100 €/year
Plant C: C_C = 53,000 €/year > $P_{CC}L_C$ = 51,300 €/year
Plant D: C_D = 81,000 €/year > $P_{DD}L_D$ = 79,200 €/year
Plant E: C_E = 144,000 €/year > $P_{EE}L_E$ = 142,500 €/year

As a consequence, we know that the resulting Business-As-Usual situation is equal to a situation in which no plant cooperates (i.e., strategic situation (NI, NI, NI, NI, NI). From Table 6.11, we can conclude that the investment cost on a multi-plant level is minimized in the all-invest situation (i.e., the To-Be situation).

After compensating all losing plants with the surpluses of all winning plants in the To-Be situation and after redistributing all additional collaboration benefits, we find the expected costs as displayed in Table 6.12.

Consequently, let us assess whether all plants are persuaded to invest in prevention measures.

Plant A: u_A (NI, NI, NI, NI, NI)* = € 18,310.19 < u_A (I, NI, NI, NI, NI) = € 21,855.99
Plant B: u_B (NI, NI, NI, NI, NI)* = € 24,483.51 < u_B (NI, I, NI, NI, NI) = €27,529.96
Plant C: u_C (NI, NI, NI, NI, NI)* = € 52,353.41 < u_C (NI, NI, I, NI, NI) = € 55,200.48

Table 6.11 Expected costs per plant in situation (4)

Strategic situation	*Plant A*	*Plant B*	*Plant C*	*Plant D*	*Plant E*	Total expected costs
(I, I, I, I, I)	20,000.00	25,000.00	53,000.00	81,000.00	144,000.00	**323,000.00**
(I, I, I, I, NI)	20,144.00	25,968.00	53,162.00	81,484.00	142,500.00	323,258.00
(I, I, I, NI, I)	20,768.00	25,836.00	53,432.00	79,200.00	145,691.00	324,927.00
(I, I, NI, I, I)	20,544.00	25,473.00	51,300.00	82,672.00	144,475.00	324,464.00
(I, NI, I, I, I)	20,400.00	23,100.00	53,729.00	81,836.00	144,304.00	323,369.00
(I, I, I, NI, NI)	20,912.00	26,803.99	53,594.00	79,683.65	144,188.46	325,182.10
(I, I, NI, I, NI)	20,688.00	26,440.99	51,461.88	83,155.99	142,974.29	324,721.15
(I, NI, I, I, NI)	20,544.00	24,067.59	53,891.00	82,320.00	142,803.54	323,626.13
(I, I, NI, NI, I)	21,311.99	26,308.99	51,731.67	80,870.80	146,165.99	326,389.45
(I, NI, I, NI, I)	21,168.00	23,935.65	54,161.00	80,035.40	145,994.99	325,295.03
(I, NI, NI, I, I)	20,944.00	23,572.80	52,028.45	83,507.99	144,779.00	324,832.23
(I, I, NI, NI, NI)	21,455.99	27,276.97	51,893.55	81,354.44	144,662.74	326,643.69
(I, NI, I, NI, NI)	21,311.99	24,903.23	54,322.99	80,519.05	144,492.00	325,549.26
(I, NI, NI, I, NI)	21,088.00	24,540.39	52,190.32	83,991.98	143,277.83	325,088.51
(I, NI, NI, NI, I)	21,711.99	24,408.44	52,460.11	81,706.18	146,469.98	326,756.71
(I, NI, NI, NI, NI)	21,855.99	25,376.01	52,621.99	82,189.82	144,966.28	327,010.09
(NI, NI, NI, NI, NI)	19,455.58	25,628.90	53,498.80	83,640.74	146,502.93	328,726.94
(NI, NI, NI, NI, I)	19,311.61	24,661.33	53,336.93	83,157.10	148,008.94	328,475.92
(NI, NI, NI, I, NI)	18,687.76	24,793.27	53,067.14	85,443.94	144,814.51	326,806.62
(NI, NI, I, NI, NI)	18,911.71	25,156.11	55,200.48	81,969.98	146,028.66	327,266.94
(NI, I, NI, NI, NI)	19,055.67	27,529.96	52,770.37	82,805.37	146,199.40	328,360.77
(NI, NI, NI, I, I)	18,543.79	23,825.69	52,905.27	84,959.95	146,317.99	326,552.69
(NI, NI, I, NI, I)	18,767.74	24,188.54	55,038.48	81,486.34	147,533.96	327,015.06
(NI, I, NI, NI, I)	18,911.71	26,561.99	52,608.50	82,321.73	147,704.96	328,108.88
(NI, NI, I, I, NI)	18,143.88	24,320.48	54,768.49	83,771.98	144,340.23	325,345.06
(NI, I, NI, I, NI)	18,287.85	26,693.99	52,338.71	84,607.96	144,510.97	326,439.48
(NI, I, I, NI, NI)	18,511.80	27,056.98	54,471.49	81,134.60	145,725.13	326,899.99
(NI, NI, I, I, I)	17,999.91	23,352.89	54,606.49	83,287.99	145,843.00	325,090.28
(NI, I, NI, I, I)	18,143.88	25,726.00	52,176.83	84,123.98	146,013.99	326,184.68
(NI, I, I, NI, I)	18,367.83	26,089.00	54,309.49	80,650.95	147,229.97	326,647.25
(NI, I, I, I, NI)	17,743.97	26,221.00	54,039.50	82,935.99	144,036.69	324,977.15
(NI, I, I, I, I)	17,600.00	25,253.00	53,877.50	82,452.00	145,539.00	324,721.50

Table 6.12 Expected costs per plant in strategic situation (NI, NI,NI, NI, NI)*

Strategic situation	*Plant A*	*Plant B*	*Plant C*	*Plant D*	*Plant E*	Total expected costs
(NI, NI, NI, NI, NI)*	18,310.19	24,483.51	52,353.41	82,495.35	145,357.54	323,000.00

Plant D: u_D (NI, NI, NI, NI, NI)* = € 82,495.35 < u_D (NI, NI, NI, I, NI) = €85,443.94
Plant E: u_E (NI, NI, NI, NI, NI)* = € 145,357.54 < u_E (NI, NI, NI, NI, I) = €148,008.94

Hence, none of the plants is persuaded to invest in prevention measures after redistributing the collaboration benefits. It is essential to note that in this situation, it is impossible by using the CIA to persuade all chemical plants without the use of external incentives.

Let us now discuss a theoretical exercise (not being part of the CIA approach) and determine the possibility to achieve an all-invest situation after all for situation (4). In order to calculate the amount of external incentives necessary to tip all chemical plants, we need to compare the expected costs in the Business-As-Usual situation (i.e., without the realization of collaboration benefits since they do not occur in this example) with the expected costs necessary to tip each plant individually.

Plant A: u_A (NI, NI, NI, NI, NI) = € 19,455.58 < u_A (I, NI, NI, NI, NI) = € 21,855.99
Plant B: u_B (NI, NI, NI, NI, NI) = € 25,628.90 < u_B (NI, I, NI, NI, NI) = € 27,529.96
Plant C: u_C (NI, NI, NI, NI, NI) = € 53,498.80 < u_C (NI, NI, I, NI, NI) = € 55,200.48
Plant D: u_D (NI, NI, NI, NI, NI) = € 83,640.74 < u_D (NI, NI, NI, I, NI) = € 85,443.94
Plant E: u_E (NI, NI, NI, NI, NI) = € 146,502.93 < u_E (NI, NI, NI, NI, I) = €148,008.94

The external incentive necessary to tip every plant individually then is:

Plant A: $S_A = T_A$ = (€ 21,855.99 – € 19,455.58) = € 2,400.41
Plant B: $S_B = T_B$ = (€ 27,529.96 – € 25,628.90) = € 1,901.06
Plant C: $S_C = T_C$ = (€ 55,200.48 – € 53,498.80) = € 1,701.68
Plant D: $S_D = T_D$ = (€ 85,443.94 – € 83,640.74) = € 1,803.20
Plant E: $S_E = T_E$ = (€ 148,008.94 – € 146,502.93) = € 1,506.01

Total amount of incentives needed from outside the chemical cluster: € 9,312.36

In this situation, the all-invest situation is not a feasible outcome since we cannot guarantee that the chemical plants obtain these incentives from outside the industrial area. The outcome according to the CIA approach is therefore a situation in which none of the plants decides to invest in prevention measures, i.e., strategic situation (NI, NI, NI, NI, NI).

6.4 Discussion of the CIA Approach

The primary goal of the CIA approach is to stimulate cooperation among a cluster of chemical plants in order to reduce the likelihood and the consequences of a domino effect occurring within the industrial area. A supra-plant body, called the MPC, stimulates cooperation through realizing collaboration benefits. We assume in this chapter that such a body, would it be created, would only have very limited resources, and that it would not have any budget at its disposal. The illustrative examples of the previous section show how, by means of the proposed Cooperative Investments Approach, collaboration with regard to domino effect prevention can still be stimulated among chemical plants. However, as Fig. 6.1 (see also Reniers et al. 2012) shows, the CIA approach has only limited power to stimulate collaboration on the one hand and realize collaboration benefits on the other hand. The MPC Data Administration's starting point for realizing collaboration benefits is the difference in expected losses between the Business-As-Usual situation (i.e., the strategic situation that would result without the use of the CIA approach) and the To-Be situation for all plants. The expected losses in the Business-As-Usual situation result from the fact whether prevention management perceives it less costly to invest in domino effect prevention rather than risking the costs and probabilities of an internal domino effect. Since prevention management cannot influence the strategy choices of the other prevention managers, they can only base their decision whether or not to invest on the difference in cost between investing (with a resulting investment cost c_i) and not investing (with a resulting implicit cost $P_{ii}L_i$).

In situation (1) it is less costly for every plant to invest in domino effect prevention rather than risking an internal domino effect. Therefore, the outcome of the CIA approach (i.e., the strategic situation that maximizes the collaboration benefits) is a situation of full cooperation. It is, however, important to note that this situation would also have been the outcome without the use of CIA. In a chemical cluster, it is evidently not guaranteed that $c_i \leq P_{ii}L_i$, and hence other situations may occur, discussed hereafter.

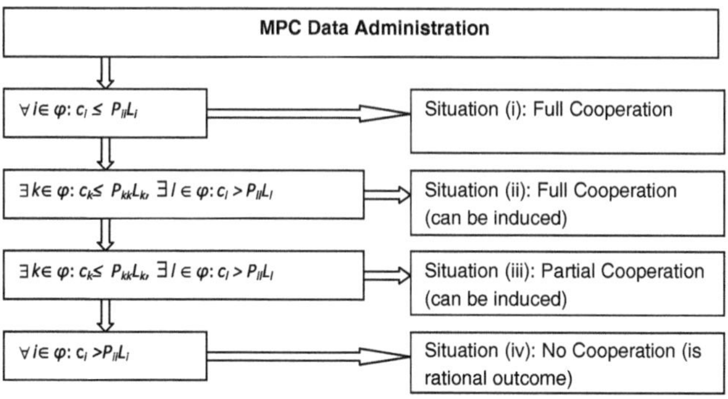

Fig. 6.1 Stepwise plan of the cooperative investments approach

Situation (2), in turn, shows that by means of the CIA full cooperation is possible even though it is not for every plant in the cluster less costly to invest in domino effect prevention. However, situation (3) shows that it is possible that full cooperation not always maximizes the collaboration benefits of the chemical cluster in a situation where it is not for all plants less costly to invest in domino effect prevention. Situation (4) shows that without the use of external incentives, the CIA approach cannot always stimulate cooperation. If it is for every plant more expensive to invest in domino effect prevention than to not invest, a situation of no cooperation is the only possible outcome in the CIA approach.

6.5 Conclusions

A Cooperative Incentives Approach is elaborated in this chapter for stimulating cooperation across possibly competing chemical plants in an industrial area. It should be stressed that some important simplifying assumptions were made in the model. First, decision makers are assumed to be risk neutral. Second, domino effects, if they occur, are assumed to have disastrous consequences, destroying entire plants. Third, prevention measures against domino effects are assumed to be fully effective and completely protect against such accidents. Obviously, these assumptions give an indication about possible future research to be carried out to lower the approach's limitations. Since we also assume that company prevention management has limited knowledge about other plants' strategies, it solely bases its cross-plant prevention decisions on their own preferences and on the perceptions they have about the preferences of other plants' prevention management. It turns out that prevention management let its decisions be directed by the simple fact whether or not it is more or less expensive to invest in domino prevention measures than to not invest in such measures and to face potential internal domino effect losses. Confidentiality issues are solved in the model by having a supra-plant body, called the Multi-plant Council, processing cross-plant information. Since the MPC disposes of all required data of every plant, it can calculate the annual expected losses for all plants in all strategic cooperative situations. By doing so, the strategic collaborative situation of the cluster minimizing the total annual investment costs, or in other words maximizing the collaboration benefits, is determined. In fact, cooperation benefits might not always lead to an *all-invest* situation. External incentives such as subsidies or taxes granted or imposed by external parties such as authorities or insurance companies may indeed be needed to shift plants from not investing toward investing.

References

Hubbart DW (2009) The failure of risk management: why it's broken and how to fix it. John Wiley and Sons, Hoboken

Reniers GLL, Cuypers S, Pavlova Y (2012) A game-theory based multi-plant collaboration model (MCM) for cross-plant prevention in a chemical cluster. J Hazard Mater 209:164–176

Chapter 7
Management Roadmap to Enhance Safety Collaboration in Chemical Industrial Parks

7.1 Introduction

Enhancing domino prevention within chemical industrial areas is actually a threefold task: first, a Multi-Plant Council (MPC) has to be established within an industrial area of chemical plants, second the MPC has to gather all the necessary information from all individual plants belonging to that multi-plant area and do the game-theoretical calculations and simulations, and third the MPC has to convince the plants investing in domino effects prevention (in order to achieve an optimal situation for the chemical cluster from a game-theoretical point of view). In industrial practice, this threefold task is much more difficult than it would seem, mainly due to hesitation by plant managers to collaborate on a strategic level. Nonetheless, the benefits of such cooperation can be substantial.

The Multi-Plant Council is thus responsible for efficiently elaborating the game-theoretical model. The model, its elaboration, and its results should be independent of the size of the multi-plant area (e.g., the number of plants, the surface area, or the amount of hazardous materials situated within the area). It should be applicable to every multi-plant situation, regardless of differences in single plant safety cultures. Model implementation guidelines need to be generic but nevertheless understandable to a variety of different clusters, situations, regulations, etc.

7.2 From Individual Plant Safety Information to Multi-Plant Safety Information

Plant safety information, the capacity to keep the chemical plant as safe as possible, remains the key to individual plant safety. However, due to the ever increasing amount of hazardous chemicals processed at cluster level and the growing public pressure on the industry to be safe, the particular safety needs required to satisfy every plant's individual safety situation are ever more dependent on the neighboring plants as well. Therefore, individual plant information on

G. Reniers and Y. Pavlova, *Using Game Theory to Improve Safety within Chemical Industrial Parks*, Springer Series in Reliability Engineering,
DOI: 10.1007/978-1-4471-5052-7_7, © Springer-Verlag London 2013

safety should be aggregated and upgraded to multi-plant information. To this end, the individual plants have to develop, collect, filter, save, distribute, and apply safety information processes at a multi-plant level. To attain this goal, harnessing, enhancing, and supporting the common information processes of a group of chemical corporations have to be done intentionally and effectively by a supra-plant organization, in this book referred to as the Multi-Plant Council. This so-called Multi-Plant Council should thus be responsible for managing collective safety information.

To efficiently manage multi-plant safety information, individual company top managements should fully recognize that sharing information on all levels (operational, tactic, and strategic), contributes to the long-term safety success of corporations, individually, as well as clustered. Around the world, the different multi-plant areas' long-term safety successes and leading edge positions will depend on their capacity for information exchange and their efficiency in handling the cross-plant information.

Optimization as regards the prevention of domino effects within an industrial area is characterized by gathering relevant information as a group of organizations by questioning the issues on individual plant level that can affect the entire group or parts of the group. It is obvious that such optimization can only be achieved by intensive cooperation and information exchanges based on solid agreements, guidelines, and procedures between plants. Knowledge management (KM) technologies may be used to this end. Inspired by KM, processes resulting in the coordination of cross-company cooperation should be elaborated. Coordination processes can be viewed upon as patterns of simple dependencies and activities that describe a mechanism for managing the relationship implied by the dependency. In order to work out these coordination processes, it is necessary to define the individual roles, processes, and technologies that can be combined in numerous ways to create unique KM solutions for a specific multi-plant area of chemical companies. This can be accomplished by setting up a Knowledge Network, e.g., the network depicted in Fig. 7.1.

By using similar networks, knowledge sharing can be developed and promoted within the multi-plant area. This, in turn, leads to efficient information sharing. By safety data sharing, existing safety practices and measures can be leveraged per plant, and safety knowledge transfer can be accelerated within the multi-plant area. Coordination processes can thus be organized by implementing game-theoretic modeling results, making the multi-plant area safety network a shared space where people can work together and by doing so, achieve step-by-step goals (e.g., guide the plants toward optimized domino effects investment strategies) resulting in continuous safety improvement cycles at multi-plant level as well as at single plant level.

To efficiently enhance collaboration within a chemical multi-plant area, in order to use game theory amongst others, requires a highly complex management design, which needs to be tailor-made for every multi-plant area. Such a management design should be drafted by the Multi-Plant Council. The next section makes powerful suggestions for this kind of design elaboration.

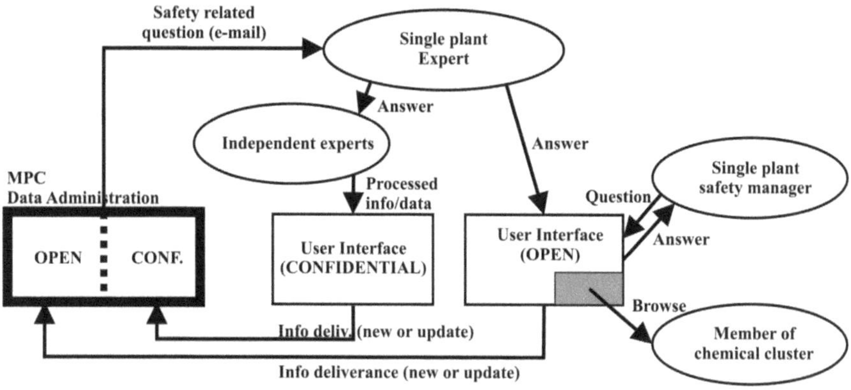

Fig. 7.1 Knowledge network system architecture for safety collaboration in a chemical multi-plant area

7.3 Toward a Design Code of Good Practice for Enhancing Safety Collaboration Within a Chemical Industrial Park

As the size and the complexity of chemical multi-plant areas grow, the recognition, identification, and adequate management of cross-company safety issues become increasingly and correspondingly important. A variety of people needs to be coordinated, a multitude of tasks needs to be carried out, responsibilities need to be identified and depending on the situation, be shared, information needs to be gathered, and, again depending on the type of information, be shared, etc. Hence, the focus of the design effort lies in streamlining and integrating the different tasks, people, resources, and information, by identifying and properly managing their interdependencies and mismatches.

The problem of connecting the tasks, people, resources, and information, is related to concepts such as resource flows, resource sharing, and timing dependencies. Malone et al. (2003) indicate that the design of associated coordination protocols involves a set of mechanisms such as shared events, invocation mechanisms, and communication protocols. Therefore, multi-plant safety collaboration enhancement can be captured in a design space that assists designers belonging to the MPC to construct a coordination process that manages a given dependency type, simply by selecting the value of a number of design dimensions. To design and implement safety collaboration within a multi-plant environment of chemical facilities, all activities related to gathering information (what information, by whom, and when), managing and processing information, drawing conclusions, and convincing single plant managers, need to be integrated. To streamline this very complicated management problem, activities need to interconnect with other activities, either because they use resources produced by other activities, or because they share resources with other activities. Moreover, the interconnection needs to be performed on three different levels: intra-company, cross-company, and supra-company.

The development of a *design code of good practice* aims to reduce the consideration of complicated interdependencies between tasks, resources, people, and information, to a routine design problem, capable of being assisted, or even automated, by computer tools in the long run.

In order to propose these design guidelines that deal with the integration of multi-plant safety responsibilities, tasks, etc., it is first imperative to identify the people, tasks, and resources involved in establishing and managing collaboration in a chemical multi-plant area. In order to establish a general guideline to address multi-plant interconnection problems to be used by a variety of companies, it is necessary to make the important choice of the generic dependency types. Three dependency types can be identified (Malone et al. 2003):

1. *Flow dependencies* encode relationships between producers and consumers of resources. Coordination protocols for managing such dependencies decompose into protocols which ensure accessibility of the resource by the consumers, usability of the resource, and also synchronization between consumers and producers.
2. *Sharing dependencies* represent relationships among consumers who use the same resource or producers who produce for the same consumers. Coordination protocols for such dependencies divide a resource among different users or enforce mutual exclusion.
3. *Timing dependencies* express constraints on the relative flow of control among a set of activities. Such dependencies are used in the decomposition of cooperation protocols for flow and sharing dependencies.

Several issues are to be considered for producing a dependency diagram, e.g., flows into a common activity may indicate a timing dependency; flows out of a common activity may indicate a sharing dependency, etc. Note that a dependency diagram represents a further abstraction and hence simplification of a flow graph, the "importance" of a dependency being relative to the point of view and judgment of the process observer. To draft this kind of diagram, all the flow, sharing, and timing dependencies which are implicit in the heuristic are drawn. Next, those dependencies that are unimportant to the observer are removed. A dependency might be considered unimportant for several reasons, e.g., its effects on the process outcomes of interest are insignificant or it has a significant effect but is easily managed. Such dependency diagrams are illustrated in Figs. 7.2 and 7.3.

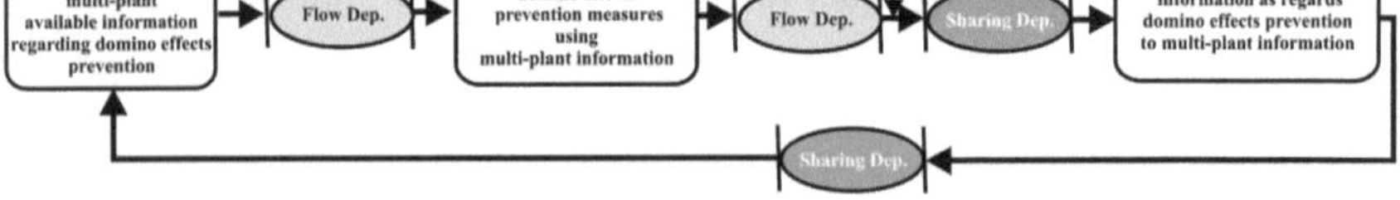

Fig. 7.2 A dependency diagram for the design cycle

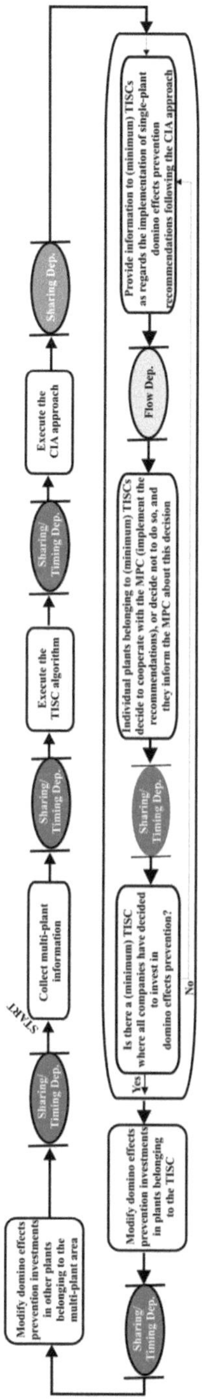

Fig. 7.3 A dependency diagram for multi-plant domino prevention improvement

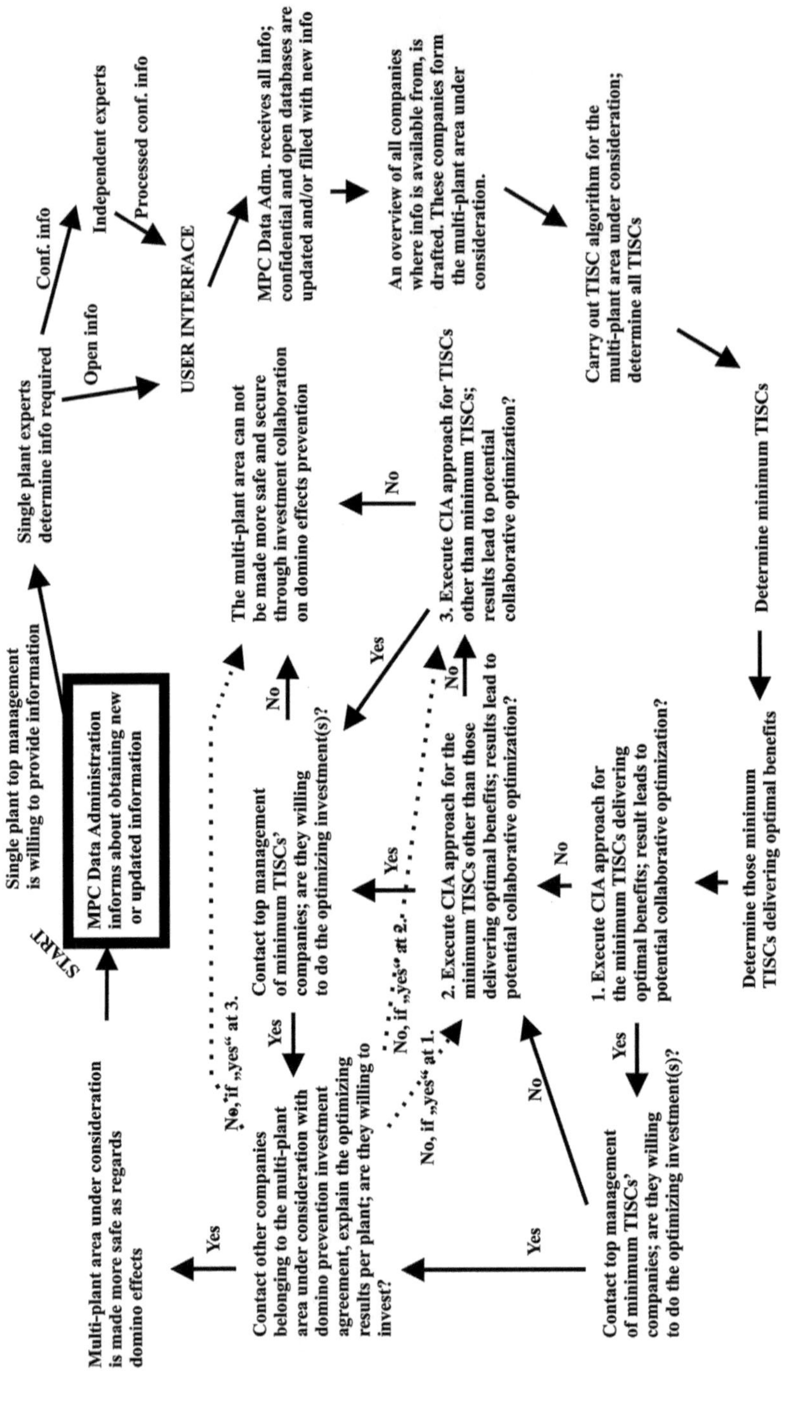

Fig. 7.4 Heuristic (flow graph) for multi-plant domino prevention investment collaboration optimization

The next step is to find a method that represents all these factors and their associations with each other in a single and easy-to-use heuristic. Figure 7.4 displays such an algorithm.

7.4 Planning for Safety Sustainability

As Beloff et al. (2005) indicate, despite the chemical industry's efforts to improve environmental and social performance and gain the public's trust, the industry still faces huge hurdles in terms of public acceptance and confidence. Fundamental topics such as historical performance, the failure of risk assessments, the intrinsic nature of hazardous chemicals, and the industry's vulnerability to terrorist threats help to explain this mistrust. In this book, a way ahead for improving safety in chemical clusters is given, by means of using game theory. This work thus constitutes important academic research to pro-actively approach the extremely important matter of preventing domino effects, on a single plant as well as on a multi-plant level. Streamlining cross-company safety collaboration further introduces new ideas in safety management practices in chemical industrial areas and enhances overall sustainability.

Sustainable safety incorporates the development and implementation of cross-company techniques and the promotion of safety to all personnel, i.e., from personnel with the lowest level of responsibility within chemical plants to personnel having a holistic view of the chemical multi-plant area. It also includes taking decisions where there is a risk of serious or irreversible harm, such as in the case of external domino accidents, even in the absence of full scientific certainty. To fully implement sustainable safety solutions in chemical multi-plant areas, multi-plant safety management should use the ORDER framework as explained in Chap. 2 in this book, and should use techniques, such as game theory, to help making the decision making more objective and more effective. The IDEAL model and the P2T model (also both explained in Chap. 2) are both models that can help in making the risk- and prevention management of industrial areas more systematic and more systemic. The following principles can be formulated to achieve chemical industrial areas that are as safe as it is economically feasible:

1. Implement the ORDER framework, the IDEAL model, and the P2T model at all levels within a chemical multi-plant area;
2. Push forward a Multi-Plant Council in every chemical cluster, which is responsible for optimizing safety within the cluster;
3. Strive to maximize safety collaboration between plants as much as possible/feasible, and use game theory to identify those individual plants and those couples/triples/etc., of plants (TISCs) that can induce domino prevention collaboration in a setting of many chemical plants situated within an industrial area.

Crucial for sustainable development in the chemical industry is the need to close the safety collaboration gap which currently exists in the industry. By implementing the game-theoretical models and approaches elaborated and explained in this book, safety within chemical multi-plant areas may be tackled much more efficiently and much more effectively as is the case at present. Since none of the individual plants has sufficient technical, regulatory, or negotiating power to force the others to behave in accordance with a common action safety framework, each company learns in the course of action how to explore its own segment of accident complexity, accumulating experience itself and at the same time benefiting from the experience of others. The result is a circuit of self-organization that, through repeated attempts, leads to the development of a safety site-integrated multi-plant area with an efficient internal organization that can respond flexibly to the demands of multi-plant members and public authorities or to the demands of the community at large.

7.5 Conclusions

A simplified dependency diagram has been proposed to manage and control safety collaboration in multi-plant industrial areas. The diagram explains the integration of the components and the mechanism of the collaboration. It helps to automate multi-plant safety design and enhances the awareness of intra- and inter-company safety aspects for optimizing single plant and multi-plant safety.

Prior to drafting a framework that encompasses and relates synchronization, communication, and resource allocation considerations, the Multi-Plant Council has to introduce a technology for coordination processes (which is multi-plant-specific). Hence, this leads to a two-step elaboration of multi-plant safety. First, the MPC is encouraged to draft multi-plant-specific coordination processes using a software/service KM-based multi-plant collaboration platform. Second, the MPC develops safety collaboration step-by-step according to the proposed dependency diagrams, and using the game-theoretical models and approaches elaborated in this book.

Reflecting on the implementation of plant and multi-plant safety, knowledge must be integral to the way in which safety managers come to make sense of continuously optimizing safety in a chemical industrial area. It will increasingly become a routine competency for making effective and (if necessary) rapid company and cross-company safety-related decisions.

References

Beloff B, Lines M, Tanzil D (eds) (2005) Transforming sustainability strategy into action. The chemical industry. Wiley, New Jersey

Malone TW, Crowston K, Herman GA (2003) Organizing business knowledge. The MIT process handbook. The MIT Press, Cambridge

Chapter 8
Conclusions and Recommendations

This book aims at supporting the expansion of the conceptual cross-plant safety framework in chemical industrial clusters. The main objectives were to resolve practical cooperation difficulties and to indicate new solutions to the old problems by giving practitioners a deeper understanding of the nature of safety incentives/taxes and cooperation.

While the first half of the book provided background information regarding safety issues on chemical industrial parks and theory of strategic cooperation (Chaps. 1 and 2), the second half of the book focused on providing an alternative viewpoint on safety problems, which gives an increased insight into the objective nature of decision making. Using mathematical modeling and applying game-theoretic analysis (Chaps. 3 and 4), we demonstrated how interdependence on operational level and, consequently, strategic level were able to play a crucial role for the resulting decisions of the plants' managers to cooperate.

The game model of strategic cooperation among the chemical plants belonging to a park, was shown to be a game of multiple equilibriums, i.e., potential outcome of strategic decision making regarding joint investments into domino accidents prevention might not be unique. There are several factors that affect the result. First of all it is important to know how each plant perceives the difference between the expected direct losses from an internally occurred accident versus individual investment costs that prevent an internal domino accident. If the costs outweigh the expected losses for all plants, the most likely result is that no cooperation regarding joint investment will be achieved. However, it does not necessarily indicate a negative situation: it may as well mean that the cluster of chemical plants has already undertaken a substantial program of enhancing its internal domino prevention and safety and the probabilities of domino accidents have become sufficiently low. Such a situation is unfortunately not very likely to take place in many real cases. If for some (if not for all) plant investment costs are lower than expected losses from an internally occurred accident, partial or even full cooperation within a cluster is possible.

G. Reniers and Y. Pavlova, *Using Game Theory to Improve Safety within Chemical Industrial Parks*, Springer Series in Reliability Engineering,
DOI: 10.1007/978-1-4471-5052-7_8, © Springer-Verlag London 2013

In order to promote and support possible cooperation, we introduced a Multi-Plant Council (MPC), grouping organization representatives from participating plants and independent delegates, to structure safety issues at a multi-plant level. Depending on the costs/probabilities/damages structure within a chemical industrial park, one of the possible stable outcomes could be 'full cooperation', when each of the chemical plants was better off by contributing to domino effect prevention, given that other plants do so as well. Despite full cooperation being the most efficient solution to the domino effect prevention investment problem, it might fail due to mutual suspect or mistrust and unwillingness to make individual commitment to invest. Indeed, if the starting point is when no one invests, the possibility of indirectly caused damage even in case one plant invests in individual safety measures, can divert a plant management toward the no investment decision. In Chap. 5 we showed that in this case there exists a subgroup of plants within a cluster such that if this group jointly invests to initiate safety investments, the other plants would follow suit, because it would be in their best financial interests. Identifying and providing incentives for such a group (or groups) of plants are proposed to be one of the tasks of the MPC. An algorithm to enhance safety collaboration within chemical clusters and inducing full cooperation was introduced and discussed in Chap. 5.

Chapter 6 widens the advisory and coordinating role of the MPC by considering a situation when only partial cooperation is a stable outcome. It took us to the settings when some, but not all of the plants, were better off by contributing to domino effect prevention given that other plants did so as well. Even in such a situation, clearly a more challenging case, it was possible to achieve a better result and get all plants to cooperate. Whether or not it was feasible depends on the total difference between individual investment costs in full cooperation and the expected losses in case of partial cooperation: some plants were more interested in cooperation than other plants, and the question was whether plants which did not benefit from investment in preventive measures, could sufficiently be compensated by those plants which were better off if full cooperation took place. As a result, it was possible to achieve full or possibly enhanced partial cooperation.

Chapters 5 and 6 suggest modeling approaches to enhance cross-plant prevention assuming that knowledge acquisition can be handled beforehand to obtain a truthful structure of cost/damages information. An interesting development of the game-theoretic modeling of cross-plant prevention in industrial clusters concerns the application of a mechanism design to motivate players to disclose their private information in games of incomplete information (Bayesian games). It can potentially be important in cases when the Multi-Plant Council would like to know the true potential damage and prevention costs of plants in the cluster but knows that it is in the plant management's interest to distort the truth. In such a case, the MPC can design a game whose rules can influence others to act the way the MPC would like. This book is mainly devoted to understanding implications of players' interdependences for analyzing prospects for joint preventive measures against domino accidents and sustaining such cooperative incentives. The role of the principal player, which would coordinate this process, has been demonstrated through the MPC. However, the role and various aspects of the MPC coordination can be put even further by applying mechanism design techniques.

In summary, game theory gives mathematical representations of chemical plants' preferences and makes predictions about investment choices and outcomes, resulting from their interaction. In practice, game theory can be considered as an especially useful tool in various strategic and/or cooperative decision-making situations in chemical industrial parks. For instance, when having to choose whether to pay for additional information and to make then a decision or to work under uncertainty, whether to follow an advice or try to imitate the behavior of other people, or whether to dictate one's own decision or discuss it with an opponent, game theory can guide a chemical plant to choose the most beneficial alternative.

As competition on the world market keeps on increasing and the clustered companies continue developing new products and production methods, it will enhance the potential for cooperative behavior within chemical industrial parks to decrease their expenditures on safety and potential losses. However, individual rationality may give a company an incentive to free-ride in order to enjoy the benefit from the collaborative rivals, exploiting so-called spillover effects. Consequently, the issue of finding a balance between collaborative and stand-alone behavior within a chemical cluster is the primary target of this book. Though the game-theoretic approach has been vastly utilized in economics of industrial organizations and also, since 9/11, it is ever more used for security problems in industrial contexts, it has hardly been used to tackle safety in multi-plant chemical industrial settings.

The intended audience for this book concerns not only risk managers and safety managers of chemical organizations, although these professions might of course benefit most from the ideas, models, and suggested approaches which are elaborated and explained. Researchers in cooperative and noncooperative game theory may also benefit from the practical insights resulting from the applications in the particular industrial sector focused upon. Furthermore, chemical companies' top management can use the book to identify their own situation, to discuss it within their company (Health, Safety, Environment and Quality department, production departments, etc.) and to act accordingly, and to improve strategic safety collaborative arrangements. In fact, in general, all process safety researchers and practitioners from industry and academia may benefit from the book. We therefore strongly recommend that they use the insights explained in the book.

MIX
Papier aus verantwortungsvollen Quellen
Paper from responsible sources
FSC® C105338

If you have any concerns about our products,
you can contact us on
ProductSafety@springernature.com

In case Publisher is established outside the EU,
the EU authorized representative is:
Springer Nature Customer Service Center GmbH
Europaplatz 3, 69115 Heidelberg, Germany

Printed by Libri Plureos GmbH
in Hamburg, Germany